# Ecotourism Impacts on Indigenous Peoples

# Ecotourism Impacts on Indigenous Peoples

Edited by
Wayne A. Babchuk and Robert K. Hitchcock

LEXINGTON BOOKS
*Lanham • Boulder • New York • London*

Published by Lexington Books
An imprint of The Rowman & Littlefield Publishing Group, Inc.
4501 Forbes Boulevard, Suite 200, Lanham, Maryland 20706
www.rowman.com

86-90 Paul Street, London EC2A 4NE



British Library Cataloguing in Publication Information Available

**Library of Congress Cataloging-in-Publication Data**

Names: Babchuk, Wayne A., editor. | Hitchcock, Robert K., editor.
Title: Ecotourism impacts on indigenous peoples / edited by Wayne A. Babchuk and Robert K. Hitchcock.
Description: Lanham, Maryland : Lexington Books, 2024. | Includes bibliographical references and index. | Summary: "This book examines the effects of ecotourism on Indigenous peoples chronicling the costs and benefits of ecotourism from a comparative and anthropological perspective"-- Provided by publisher.
Identifiers: LCCN 2024028523 (print) | LCCN 2024028524 (ebook) | ISBN 9781793643841 (cloth) | ISBN 9781793643858 (epub)
Subjects: LCSH: Ecotourism--Social aspects. | Tourism--Anthropological aspects. | Culture and tourism. | Indigenous peoples--Ethnic identity.
Classification: LCC G156.5.E26 E35394 2024 (print) | LCC G156.5.E26 (ebook) | DDC 305.8--dc23/eng/20240718
LC record available at https://lccn.loc.gov/2024028523
LC ebook record available at https://lccn.loc.gov/2024028524

∞™ The paper used in this publication meets the minimum requirements of American National Standard for Information Sciences—Permanence of Paper for Printed Library Materials, ANSI/NISO Z39.48-1992.

# Contents

# List of Illustrations and Tables

## FIGURES

## TABLES

# Acknowledgments

Support for the editing of this book has been provided by Lexington Books. We wish to thank Kasey Beduhn of Lexington Books for approaching us about the possibility of turning a symposium held at the 77th Annual Meetings of the Society for Applied Anthropology held in Santa Fe from March 27 to March 31, 2017, into this volume. Alexandra Rallo of Lexington Books provided us with substantial help in the production of this volume. We wish to thank Gomathy Ilammathe, Megan White, Emily Kasprzyk, and Ryan Dradzinski for helping us enormously with the finalization of this volume. We also thank Debbie Miller and Melinda C. Kelly for their editorial assistance. In addition, we wish to express our appreciation to the University of Nebraska-Lincoln and the University of New Mexico for their support during the course of the writing and editing process.

*Chapter 1*

# Ecotourism, Indigenous Peoples, and Societal Impacts

Wayne A. Babchuk and Robert K. Hitchcock

The International Ecotourism Society (2018:1) defines ecotourism as "responsible travel to natural areas that conserves the environment and improves the well-being of local people." Ecotourism is promoted by international organizations such as the World Bank (2018, 2020a, b) and the United Nations World Tourism Organization (UNWTO) as a useful way to enhance livelihoods, incomes, and well-being of both countries and local communities and to alleviate poverty. As the UNWTO notes, "The tourism sector is committed to halting and reversing biodiversity loss while promoting environmental conservation and restoration. It seeks to enhance people's well-being, livelihoods and economies, leveraging tourism's unique potential as an agent for positive change" (UNWTO 1999:3).

Tourism consists of traveling from one location to another, either within the country's borders or across borders, for leisure, educational, or other reasons, usually for a short period of time. Ecotourism travel is often scheduled for a week or two at most and, as a result, is seen by some as having a lower degree of impact on the host populations.

Ecotourism is described by a number of different terms: green tourism, soft tourism, pro-poor tourism, responsible tourism, alternative tourism, and sustainable tourism (Ceballos-Lascurain 1992; Honey 1999; Stronza, Hunt, and Fitzgerald 2020:231). Standard tourism has been seen as "the good, the bad, and the ugly" (Rosenow and Pulsipher 1979). It has also been characterized by Crick (1984) as being represented by "the 5 S's": sun, sex, sights, savings, and servility. As Boo (1990) notes, however, ecotourism has both potentials and pitfalls.

Until 2020, tourism was one of the largest contributors to the world economy and was the world's largest industry. The number of tourists traveling internationally declined substantially in 2020–2021 as the COVID-19 pandemic saw many of the world's countries close their borders. The war in Ukraine that began in February 2022, and together with global inflation, contributed to a decline in international travel. It is likely that as the global health and economic situations improve in the coming years, tourism will expand again as it did from 1950 to 2010.

The impacts of ecotourism on Indigenous people have been a long-standing theme in the literature on tourism (Butler and Hinch 1996; Hinch 1998; Honey and Thullen 2003; Aicken and Ryan 2005; Johnston 2006; Zeppel 2006). Part of the reason for this is that Indigenous peoples have been considered to be highly vulnerable to visits of tourists. Ecotourism is sometimes represented as being beneficial to Indigenous people because it provides them with markets for their goods and services, and it is said to promote conservation, socioeconomic development, and empowerment.

Some Indigenous support organizations, such as Survival International in London, have waged long-term campaigns against tourism because of its perceived negative effects on Indigenous peoples (see Survival International 1994). Survival International sees tourism not only as intrusive but also as exploitative and culturally destructive. Tourism has been used by some governments and organizations as a justification for the removal of Indigenous peoples from their ancestral territories (see, for example, Sapignoli 2018, who discussed the case of the Central Kalahari Game Reserve in Botswana).

Indigenous peoples, sometimes described as native peoples, aboriginal peoples, traditional peoples, or tribal peoples, are found in some 90 of the world's 193 countries, and number at least 900,000,000 people (Daes 2008; Mamo 2023). A sizable proportion of these groups exist in the world's conservation hotspots, which are of considerable ecotourist interest. These same places are often vulnerable to overutilization if the tourist density is too high.

As Stronza, Hunt, and Fitzgerald (2020) note, there is a serious need for balanced analyses of ecotourism that consider both social and environmental impacts equally. Community-based ecotourism (CBET) is being practiced increasingly in various parts of the world, and there are cases such as those in Namibia where success has been achieved. There has been an expansion of participatory evaluations of CBET, but generally they continue to be relatively uncommon. Diachronic analysis of cases is needed, along the lines of those done in the Serengeti ecosystem in East Africa (Sinclair et al 2015) and the Okavango Delta region, a World Heritage Site (WHS) in northwestern Botswana (Mbaiwa 2003, 2012; Mbaiwa and Stronza 2010). Hundreds of Khwe San were removed from the Moremi Game Reserve in the Okavango in the 1970s in order to promote tourist use (Mbaiwa 2016; Maitseo Bolaane, personal communication, January 2023). Today, there are over 180 tourist camps in the Okavango, only a small proportion of which provide employment for Indigenous people.

Many nation-states, including Bolivia, Botswana, Costa Rica, Nepal, and Vietnam, have pushed for the expansion of ecotourism, seeing it as a way of generating income and expanding awareness of natural, cultural, and historical sites and local-level diversity. Other countries, such as Bhutan and Saudi Arabia, have been cautious about promoting ecotourism.

The costs and risks of ecotourism include exploitation, manipulation, dependency, destruction of authenticity, cultural commodification, exacerbation of gender differences, decline of food production, overexploitation of local resources, spread of disease, and expanded drug and alcohol addiction at the local level.

There are also benefits of ecotourism, including increased personal income, employment, education, increased understanding and appreciation of other people, expansion of markets for Indigenous products, sharing of Indigenous knowledge, and provision of access to information both to tourists and Indigenous people. Ecotourism also provides support for Indigenous identity.

Looked at from the perspective of Indigenous peoples, ecotourism can provide jobs, income, and markets for local products. Ecotourism also raises the issues of the politics of belonging and identity formation. Ecotourism often has costs, including undesirable environmental impacts, exacerbation of social tensions, increases of local inequality, bringing about changes in health and well-being at the local level, privileging males over females in some types of ecotourism, and contributing to increases in anti-tourist activism. Indigenous peoples around the world have argued for greater local control over ecotourism in which they can participate fully and receive both direct and indirect benefits for themselves and their children. Indigenous peoples are hoping that the ethics of ecotourism will be developed in such a way that their needs are taken into consideration (see the UNWTO's 1990 statement on ethics for the UN position).

This book seeks to examine the social, economic, environmental, and political impacts of ecotourism. Travel, as Steves (2021) notes, is a political act. Indigenous people in many parts of the world want to have greater power over ecotourism and over their own lands. This book consists of ten chapters on Indigenous peoples in Africa, Asia, South America, and North America who have been affected by ecotourism. Some of the chapters are comparative in nature, while others focus on ecotourism at the national or local levels.

Chapter 2, by Sonia Arellano-López, focuses on sustainable tourism and the marketing of Indigenous identity in Bolivia. Bolivia, as a plurinational state, has the highest percentage of Indigenous people in a nation-state in South America. The election of Evo Morales, Bolivia's first Indigenous president, in 2005, and its 2009 constitutional reform transformed the country's approach to tourism development. The presidency of Evo Morales (from 2005 to 2019) set the stage for new approaches to tourism development in Bolivia. Innovative policies for sustainable development were put in place. Indigenous peoples' role in tourism changed from being props for cultural depictions of national identity to playing a greater role than was the case in the past. Indigenous peoples in Bolivia became more business-oriented and increasingly have been able to lead economic development alongside environmental and cultural heritage management. Community-based tourism has sought to provide social and economic benefits to local communities throughout the country, and in many ways has become a leader in Indigenous, locally based tourism.

Chapter 3, by Robert Hitchcock, Melinda Kelly, and Wayne Babchuck, focuses on ecotourism impacts in three countries of southern Africa: Botswana, Namibia, and Zimbabwe. While ecotourism has been seen as a viable way to improve the well-being of the San and other Indigenous people in southern Africa, evidence suggests that this development strategy has had variable impacts. In Botswana, the promotion of tourism has resulted in requiring local people to resettle out of game reserves and WHSs such as the Central Kalahari Game Reserve and the Tsodilo Hills. In Namibia, the establishment of high-end ecotourism operations has led to the employment of key members of Nyae Nyae Ju/'hoan San society who are well-known healers, trance dancers, and craft producers being taken to places far from their homes to provide services for tourists. This type of appropriation is also the case in Botswana. In Zimbabwe, much of the tourism is either wildlife-related or cultural heritage tourism, including visiting rock art and archaeological sites such as those in the Matopos Hills, or other well-known archaeological sites. There are few or no benefits made available from tourism to the San who reside or once resided in these areas. The variable costs and benefits of ecotourism in the three countries are described in detail.

Chapter 4, by Megan Biesele, titled "Ecotourism, Anthropology, and San Community Docents and Activists: Some Interactions," reports on several illuminating examples of interactions among (1) San community docents and activists in Namibia, Botswana, and South Africa; (2) the tourists they task themselves to educate and influence; and (3) anthropologists/activists and their publications on the communities in question. The examples include, first, an ecotourism course given jointly by the Working Group of Indigenous Minorities in Southern Africa (WIMSA) and The University Centre for Studies in Namibia (TUCSIN) for San community docents held in Windhoek, Namibia, in 2007. Second, it covers sets of values guidelines created by San activists in 2012 to inform a number of San community development and ecotourism projects in Botswana and Namibia. Third, it describes recent (2015–2020) updates to docent education for ecotours at !Khwa ttu San

Education and Cultural Centre, Yzerfontein, South Africa. The examples share both increasing use and increasing critique by San docents and activists of written resources on the relevant communities as furnished by anthropologists. They suggest ways San community members may continue to increase their control of community representation and decision-making.

Chapter 5, by Fleming Puckett, examines ecotourism and cultural activities among the ǂKhomani San communities in South Africa today. Some of the best San-run tourism operations are found in South Africa. Ample research indicates the potential of San-run tourism operations to benefit San communities, including providing a platform for the practice and intergenerational dissemination of identity-affirming aspects of traditional hunting-and-gathering culture. This chapter outlines the key players, processes, and promises of the ǂKhomani San hunting operation on Erin Farm in South Africa's Kalahari Desert. The work of devoted supporters and advisers, government development funds, private loans, committed ǂKhomani leadership, and skilled, dedicated San employees—along with an absence of government-imposed hierarchical leadership structures since 2008—have combined to produce what appears to be southern Africa's most successful San-run commercial hunting venture. Individuals have come together to host and guide tourists, earn money, practice and showcase their traditional knowledge, reconnect with their lands, share the meat, delight in their emerging identity as a ǂKhomani San community, and defend the asset that Erin Farm has become.

Chapter 6, "The Mayanization of the Ch'orti': Indigenous Identity amid Tourism-based Development Initiatives in Copán, Honduras," by Fredy Rodriguez-Mejia, discusses how Indigenous people in Honduras became increasingly involved in tourism. After the political movement of the Ch'orti' Maya of Honduras gained national and international attention in the late 1990s, Indigenous people were gradually involved in tourism-based development initiatives. These initiatives created different opportunities for people to mobilize their identities in creative ways. This chapter considers the mobilization of Indigenous identity as both a livelihood strategy and a political stance. It pays particular attention to how the notion of being Maya is understood and used by non-Indigenous entrepreneurs and Indigenous people who are part of these tourism development initiatives. By relying on the works of anthropologists and archaeologists, tourism entrepreneurs are able to construct particular kinds of identities that Indigenous people are encouraged to embrace. Thus, non-Indigenous entrepreneurs can sell images of the Ch'orti' Maya (through theater performances) that create expectations among tourists about how a Ch'orti' Maya person should look, speak, or behave. At the same time, these strategies are used by Indigenous actors who formerly worked for non-Indigenous entrepreneurs in order to take advantage of their own position as Ch'orti' Maya. This chapter highlights how Indigenous people creatively use their identities and political activism to both contest and compete with non-Indigenous society in the commoditization of identity.

Chapter 7, "Aboriginal Alternative Tourism in Arnhem Land: Tourism as Cultural Learning," by Sachiko Kubota, focuses on the new tourism styles found in Arnhem Land, Australia, involving Aboriginals, specifically the Yolngu. It addresses cultural learning among the Yolngu Aboriginals and places emphasis on what aboriginal hunter-gatherers deal with changes in terms of infrastructure and technology. The chapter demonstrates that hunter-gatherer groups are not only passive recipients or victims of consequential infrastructural changes but rather demonstrate remarkable abilities, resilience, and technological ingenuity.

Chapter 8, "How Does Cultural Tourism in a National Monument Affect Its Affiliated Pueblo Tribes?" by Vidal Gonzales, Robert Hitchcock, and Melinda Kelly, examines the impacts that Bandelier National Monument, one of the premier monuments in the Southwestern United States, has on the area's affiliated Pueblo tribes. Established in 1916, Bandelier contains thousands of Ancestral Puebloan archaeological and historic sites which have been the subject of archaeological inquiry since the 1880s. Currently, the monument covers 3,677 acres (136.29 $km^2$ or 46.35 $mi^2$), but there are plans to expand it. The descendants of the Ancestral Puebloan people who occupied Bandelier between AD 1100 and AD 1550 continue to reside in the region, including those in Cochiti, San Felipe, Kewa (Santo Domingo), and Zia Pueblos, who are Keres speakers, and those in Okhay Owingeh (San Juan), Santa Clara, and San Ildefonso who are Tewa speakers. Bandelier is one of the most highly visited national monuments in the Southwestern United States, and it has had a long history of interactions with neighboring Pueblo communities. A new management strategy established in 2016 aims to include affiliated Pueblo tribes in the handling of their own cultural heritage known as the Bandelier Preservation Corps (BPC), a powerful consultation and collaborative mechanism between the monument, its employees, and the Pueblos. The BPC provides training to Pueblo youth, who learn about the cultural heritage of Bandelier National Monument and can get heritage management and cultural resource preservation experience. Their interaction with visitors ensures a process whereby the visitors can become more understanding and empathetic toward the worldview and concerns of Indigenous peoples.

Chapter 9, titled "COVID-19 and the Limits of Community-Based Ecotourism as a Sustainable Livelihood Diversification Strategy: The Case of the Indigenous Karo of Batu Katak, North Sumatra, Indonesia" by Roderick T. J. Buiskool and Stasja Koot, contributes to the growing body of literature about the effects of the COVID-19 pandemic on tourism. Buiskool and Koot describe how this "shock" became a structural "stress" for the Indigenous Karo Batak of Batu Katak living adjacent to the Mount Leuser National Park (MLNP), Sumatra, Indonesia. Many Karo in Batu Katak are involved in CBET in an attempt to diversify their livelihood opportunities, making ecotourism an important strategy in addition to mostly agricultural livelihood activities. However, the effects of COVID-19 in 2020 led to an enormous reduction in the number of tourists globally, including in Batu Katak and on the Karo specifically. The vulnerability of ecotourism for Indigenous groups as a sustainable livelihood diversification strategy is seen against the backdrop of the COVID-19 pandemic. While CBET is often considered a market-based mechanism that reduces marginalized groups' vulnerability, the authors argue that the ongoing COVID-19 pandemic has highlighted the vulnerability of CBET as a mechanism in itself. This has important implications for the future of global efforts to preserve biodiversity and empower Indigenous people through tourism: whereas ecotourism is often presented as an important livelihood diversification strategy to increase people's social sustainability, a shock/stress like the COVID-19 pandemic shows that it is important to keep approaching it as a *diversification*, and not as the dominant strategy. However, due to various attractive benefits, ecotourism can have the tendency to prevail as a livelihood strategy, which tends to increase dependency on, and vulnerability through, CBET for Indigenous and other marginalized groups, as was shown in Batu Katak and its surroundings after COVID-19.

Chapter 10, "Change among the Wannilaeto of Sri Lanka: Land, Livelihoods, and Ecotourism Impacts" by Wiveca Stegeborn, discusses the Wannilaeto (Veddahs) Indigenous people of the country of Sri Lanka. Like hunters and gatherers in many parts of the world, global, national, and local processes have affected their livelihoods and well-being significantly, particularly since Sri Lanka's independence in 1948. In the 1950s, settlers who engaged in commercial agriculture were brought to the Wannilaeto area. Rainforest land was bought, trees were logged and teak was planted for the export of precious wood. The plantations follow the eastern border of the Maduru Oya National Park. Trees and non-timber forestry products were exploited intensively by the incomers. This chapter addresses the challenges that have faced the Wannilaeto, including the implementation of the Accelerated Mahaweli Development Scheme, a large-scale agricultural irrigation project (1983–1992), which exacerbated competition for land, including traditional forestland inhabited by Wannilaeto. The Maduru Oya National Park, a protected area, was established in 1983 on Wannilaeto land. Government troops moved in for military training during the latest Sri Lankan Civil War (1983–2009) and the Wannilaeto were displaced to government Rehabilitation Villages to become rice farmers. The coronavirus pandemic has affected ecotourism in Sri Lanka substantially, and the downturn in the economy and political corruption have impacted the Wannilaeto. The Wannilaeto today are coping with tourists, human trafficking, health challenges, and incorporation into the national economy as poorly paid workers. The forest dwellers, whose forest home was integral to their hunting-gathering economy, were considered intruders in their environment, and their lands are now largely in the hands of other people. This chapter examines the varied impacts of ecotourism on the Wannilaeto.

Considering that there are as many as 900,000,000 Indigenous people residing in at least 90 of the world's 193 nation-states today (see Mamo 2023), it is important to examine the activities, including ecotourism, that affect their livelihoods, identities, and well-being. Table 1.1 presents data on the numbers of Indigenous people discussed in this volume.

Indigenous peoples are all too often seen as "exotic" or "the other" and are treated accordingly by ecotourists and ecotourism companies, as well as by the nation-states in which they reside. Sometimes Indigenous groups are pressured to dress in "traditional" clothing and to engage in staged performances such as dances and healing ceremonies that they feel are non-authentic. In some cases, such as among Pueblos in the Southwestern United States, Indigenous groups prefer not to have tourists observe their religious ceremonies. At the same time, the presence of ecotourists in some cases has had the effect of revitalizing cultural traditions and educating the young about important aspects of their cultural heritage. Ecotourism often favors elite members of Indigenous groups (Bodley 2015:167) and can have outsized impacts, particularly on remote Indigenous communities such as those in the Amazon Basin, islands of the Caribbean, and the Pacific and Indian Oceans. Ecotourism has been used in some cases to justify the removal of

**Table 1.1 Indigenous People Covered in This Volume**

| *Country* | *Group(s)* | *Number of People* |
|---|---|---|
| Bolivia | 36 indigenous peoples, including Aymara, Quechua, and Guarani | 6,016,026 51% of the population |
| Botswana, Namibia, Zimbabwe | San of various groups | 109,300 San |
| South Africa | Khwa ttu San Culture and Education Centre | 35 employees, 28 of them indigenous |
| South Africa | ǂKhomani San | 1,600 |
| United States | Pueblo Peoples in New Mexico | 9,973 in affiliated Pueblos |
| Honduras | Ch'ori' Maya | 34,000 |
| Australia | Yolnu | 5,000 |
| Indonesia | Karo | 1,100,000 |
| Sri Lanka | Wannilaeto | 2,500 |

*Source*: Authors.

Indigenous people from their ancestral lands (Brockington 2002; Igoe and Brockington 2006). It is important to build on the lessons of ecotourism to provide policy recommendations on how best to ensure high levels of local participation and to minimize negative social, economic, political, and environmental impacts.

The International Ecotourism Society (TIES) (2018) maintains that those who participate in ecotourism activities should follow these principles:

- Minimize impact.
- Build environmental and cultural awareness and respect.
- Provide positive experiences for both visitors and hosts.
- Provide direct financial benefits for conservation and development.
- Provide support for empowering local people.

As outlined in this volume, Indigenous peoples are seeking a greater role in the management of ecotourism and cultural tourism so that sustainability and community empowerment can be promoted more effectively.

## REFERENCES

Aicken, Michelle and Chris Ryan. *Indigenous Tourism: The Commodification and Management of Culture* (Amsterdam: Elsevier, 2015).

Bodley, John H. *Victims of Progress*. Sixth Edition (Lanham, MD: Rowman and Littlefield, 2016).

Boo, Elizabeth. *Ecotourism: The Potentials and Pitfalls* (Washington, DC: World Wildlife Fund, 1990).

Brockington, Dan. *Fortress Conservation: The Preservation of the Mkomazi Game Reserve, Tanzania*. (London: James Currey, 2002).

Butler, Richard and Thomas Hinch, eds. *Tourism and Indigenous Peoples* (London: International Thompson Business Press and New York: Routledge, 1996).

Ceballos-Lascurain, H. *Tourism, Ecotourism and Protected Areas* (Caracas, VE: IUCN World Conservation Union, 1992).

Crick, Malcolm. "Representations of International Tourism in the Social Sciences: Sun, Sex, Sights, Savings, and Servility." *Annual Review of Anthropology* 18 (1989): 307–344.

Daes, Erica-Irene. *Indigenous Peoples: Keepers of Our Past, Custodians of Our Future* (Copenhagen: International Work Group for Indigenous Affairs. 2008).

Hinch, T. Ecotourists and Indigenous Hosts: Diverging Views on their Relationship with Nature." *Current Issues in Tourism* 1 (1998): 120–124.

Honey, Martha. *Ecotourism and Sustainable Development: Who Owns Paradise?* Second Edition (Covelo CA: Island Press, 1999).

Honey, M. and S. Thullen, eds. *Rights and Responsibilities. A Compilation of Codes of Conduct for Tourism and Indigenous and Local Communities* (Washington,

DC: Center on Ecotourism and Sustainable Development and The International Ecotourism Society, 2003).

Igoe, Jim and Daniel Brockington. "Eviction for Conservation: A Global Overview." *Conservation and Society* 2(2) (2006): 411–432.

International Ecotourism Society. *What is Ecotourism?* (Washington, DC: The International Ecotourism Society, 2018).

Johnston, Alison M. *Is the Sacred for Sale? Tourism and Indigenous Peoples* (London and Sterling, VA: Earthscan, 2006)

Mamo, Dwayne, ed. *The Indigenous World 2023* (Copenhagen: International Work Group for Indigenous Affairs, 2023).

Mbaiwa, Joseph E. "The Socio-economic and Environmental Impacts of Tourism Development on the Okavango Delta, North-western Botswana." *Journal of Arid Environments* 54(2) (2003): 447–467.

Mbaiwa, Joseph E. "Hosts and Guests: Stereotypes and Myths of International Tourism in the Okavango Delta." In *African Hosts and their Guests: Cultural Dynamics of Tourism,* edited by Walter van Beek and Annette Schmidt, 117–136 (London: James Currey, 2012).

Mbaiwa, Joseph. "Tourism Development, Dispossession, and Displacement of the Communities in the Okavango Delta, Botswana." In *Political Ecology of Tourism,* edited by Sanjay Nepal and Jarkko Saarinen (193–206) (London: Routledge, 2015).

Mbaiwa, Joseph E. and Amanda Stronza. "The Effects of Tourism Development on Rural Livelihoods in the Okavango Delta." *Journal of Sustainable Tourism* 18(5) (2010): 635–656

Rosenow, John E. and Gerreld L. Pulsipher. *Tourism: The Good, the Bad, and the Ugly* (Lincoln, NE: Dill Publishing, 1979).

Sapignoli, Maria. *Hunting Justice: Displacement, Law, and Activism in the Kalahari* (Cambridge: Cambridge University Press, 2018).

Sinclair, A.R.E., K.L. Metzger, S.A.R. Mduma, and J.M. Fryxell. *Serengeti IV: Sustaining Biodiversity in a Coupled Human-Natural System* (Chicago and London: University of Chicago Press, 2015).

Steves, Rick. *Travel as a Political Act: How to Leave Your Baggage Behind.* Third Edition (Berkeley, CA: Avalon Travel, 2021).

Survival International. *Tourism and Tribal Peoples: The "New Imperialism*" (London: Survival International, 1994).

United Nations World Tourism Organization (1999) *Global Code of Ethics for Tourism* (Madrid: World Tourism Organization, 1999).

World Bank. *Supporting Sustainable Livelihoods through Wildlife Tourism* (Washington, DC: World Bank Group, 2018).

World Bank. *Tools and Resources for Nature Based Tourism* (Washington, DC: The World Bank, 2020a).

World Bank. *Banking on Protected Areas: Promoting Sustainable Protected Area Tourism to Benefit Local Communities* (Washington, DC: The World Bank, 2020b).

Zeppel, H.D. *Indigenous Ecotourism: Sustainable Development and Management* (Wallingford, UK: CABI International, 2006).

# *Chapter 2*

# Sustainable Tourism

## *Marketing Indigenous Identity in the Plurinational State of Bolivia*

F. Sonia Arellano-López

The[1] term sustainable tourism began to appear regularly in English-language academic literature in the early 1990s. In 1993, the specialized peer-reviewed *Journal of Sustainable Tourism* (JOST) was established to encourage discussion and research on the environmental impacts that the accelerated tourism development was causing. Since the 1990s, more than 6,000 individual academic publications, crossing a wide range of economic, social, cultural, and natural science disciplines, have been published on topics related to sustainable tourism.[2] This growth in academic interest reflects the expansion of the tourism industry as a manifestation of globalization and concern with issues of sustainable development.

Globalization, in turn, is the outcome of the rapid expansion of capitalist relations of production in developing countries, and the way in which the spread of tourism has led to destinations, local cultures and environments (such as national parks, wildlife, flora and fauna, and so on) being transformed into commodities to be consumed by tourists.[3] This type of environmental and cultural commodification process became subtler and superficially more benevolent when Agenda 21 introduced the concept of sustainable development to broad audiences. This extensive blueprint for implementing sustainable development, agreed upon by the nations at the 1992 Rio Summit, in Brazil, and reaffirmed at the World Summit on Sustainable Development (WSSD), held in Johannesburg, South Africa in 2002, influenced thinking and practice in the tourism industry. Tourism was specifically mentioned as offering sustainable development potential. It could stimulate economic growth and create business opportunities and employment. Largely, tourism could act as a catalyst for environmental and cultural protection as social and economic development processes unfolded.[4]

In 1999 the United Nations Commission on Sustainable Development (CSD) called on governments and international institutions to maximize the potential of tourism for eradicating poverty, by developing appropriate policy strategies in cooperation with governments, businesses and civil society organizations. The CSD argued that, if well designed and managed, tourism can contribute to generating decent employment and creating wealth. CSD also held that tourism can help preserve the environmental and cultural heritage assets upon which countries depend, empower host communities, generate trade opportunities, and foster peace and intercultural understanding. The same year the World Tourism Organization (WTO) adopted the *Global Code of Ethics* as a contribution to the fight against poverty. Thus, tourism became a vital sustainable development contributor (Mowforth & Munt, 2009:32–38).

The inherent relation between sustainable development and tourism produced, unlike mainstream or conventional mass-scale tourism, a new type of tourism defined by its relatively small scale and the application of prefixes such as "low impact," "responsible," "alternative," "appropriate," and "sustainable." This new type of tourism is considered to promote minimal environmental impact and high social and economic inclusion, particularly of populations that were left at the margins of previous social and economic development processes. Thus, it seeks to penetrate the less visited parts of developing countries and commodify what is there. But, as part of a sustainable development agenda, it must be built by, through and with the commitment of local communities. As Mowforth and Munt (2009:120) argued, in the field of tourism the requirements of sustainable development cannot merely be imposed; active participation by local communities is needed. Therefore, in new forms of tourism, local communities and Indigenous people, in particular, have become rights holders of their own environmental and cultural assets. They play a critical role for sustainable development and tourism. Meanwhile, politically and economically dominant stakeholders (i.e., governments, international institutions and non-government organizations, etc.) provide a particular political framework for tourism to fit the larger international sustainable development agenda (Duffy, 2002; Hall et al. 2015; Wearing & Neil, 2014).

The inter-twined issues of sustainable development and tourism outline the analytical framework for studying how this new type of tourism has become the standard model for rationalizing the construction of policies and programs intended to enhance Indigenous national identity, and political and economic participation in the Bolivian government. Prior to the election of the first Indigenous president, in 2005, and the approval of a new National Constitution, in 2009, the country's tourism institutions and plans depended strongly on initiatives taken by the private sector. Regional and local authorities had little to do with tourism planning, development, and management, other than providing operational licenses. Local communities were absent from discussions of tourism development issues, other than as a source of labor. Indigenous peoples were used as props for the cultural depiction of national identity and political cohesion. With the legal backing of the new political Constitution, the government designed and launched new economic and social development policies and programs such as, the Indigenous Autonomy Law, the Mother Earth Law and drafted the "Bolivia worthy, sovereign, productive and democratic in order to live well" socioeconomic development plan. The political Constitution, the new laws and the socioeconomic development plan became pillars for designing, planning and implementing the country's tourism agenda. Though, conventional tourism models were not completely expunged, the new type of tourism initiatives, particularly community-based tourism (CBT) and ecotourism were actively promoted. This new type of tourism reflected and contributed to internationally accepted sustainable development policies, and viewed Indigenous people as business-oriented and able to lead economic development in addition to being the stewards of the natural environment and traditional cultures. Within the country, this new type of tourism suited the government's political development discourse, an Indigenous country where a "productive communitarian socialist and economic" model has been promoted (Arce Catacora, 2011).

## SETTING THE BASE FOR AN INDIGENOUS SUSTAINABLE TOURISM

With the election of the Indigenous leader Evo Morales Ayma as Bolivia's President, in 2005, Indigenous identity became the major definer of peoples' relationship with the state and associated development planning. This relationship became more formalized with the approval of a new Constitution in 2009.[5] The Constitution established that Indigenous populations are those settled prior to the time of conquest or colonization and that are comprised within the present state boundaries. They have history, organization, language or dialect and other cultural characteristics which their members identify with acknowledging themselves as pertaining to the same social and cultural unit. They retain a territorial bond in terms of managing their habitat and their social, economic, political, and cultural institutions within the framework of the Plurinational Bolivian State. Based on this definition the Constitution recognized thirty-six Indigenous groups. According to the 2012 national census, 41 percent of the total population of approximately eleven million identify themselves as of Indigenous origin. The largest Indigenous groups—Aymara and Quechua, are settled largely in the country's Andean region; the other Indigenous groups are culturally diverse and dispersed across Bolivia's lowland region (Instituto Nacional de Estadística 2015; Sturtevant 2023).[6]

Indigenous people's predominance in the country's political arena is suggested in the national economic and social development plan "Bolivia worthy, sovereign, productive and democratic in order to live well." The plan calls for the formerly invisible and poor Indigenous peoples of Bolivia not only to be seen and heard, but to be the principal protagonists and beneficiaries of economic, political and social development policies. The plan stresses the notion of Indigenous communitarian forms of living together, respecting intercultural differences and redistributing economic and political power. It promotes mechanisms that guarantee the participation of Indigenous people in the processes of sustainable development and use of natural resources, taking into consideration the Indigenous communities' specific economic, social and cultural vision and needs. Likewise, the government's political discourse underlines that "Bolivia worthy, sovereign, productive and democratic in order to live well" is not only the practice of sustainable development policies, but a political and economic alternative against conventional western development policies. It is the production system that combines Indigenous traditional knowledge and practice regarding the environment, bringing back communitarian Indigenous production practices, which in turn would deliver harmonic and conservationist relations with the environment (Postero 2017:95–97; Galindo, 2018:18–19).

According to Article 337 of the new Constitution, tourism is a national strategic economic activity and it needs to be developed in a sustainable manner to preserve the country's cultural wealth and environmental diversity. Since ninety percent of tourism attractions are located in rural areas, tourism development should, on the one hand, create employment and generate revenues to finance development (e.g., road infrastructure, education and health programs) in the economically poorer regions. On the other hand, tourism development should serve as a mechanism to add value to the cultural Indigenous inheritance of the country. Thus, the government's responsibility is to outline a tourism agenda that emphasizes social and economic inclusion of indigenous peoples and nations of urban and rural communities that have been economically dispossessed and socially excluded. Based on the Constitution, the government included tourism as a pivotal strategy in the economic development plan. The government asserted that tourism is what will lead Indigenous people to plan, manage, control, and demand accountability for the use of their own environmental and cultural resources. In so doing, they would also validate their territorial claims, raise their incomes, value their culture and strengthen their own ancestral government institutions (UDAPE Bolivia, 2015: 6–7; Ministerio de Culturas y Turismo, 2016: 43–44).

The Tourism Law 292 "Bolivia awaits you," enacted by the government in 2012, establishes that the country's tourism policies and planning should promote and develop public, private, and CBT activities, highlighting private tourism and CBT. Private tourism refers to individuals who have the economic and social competence to develop tourism ventures. Private tourism operations are required to comply with the national sustainable development goals. CBT refers to any investment made by urban and rural communities, nations and Indigenous peoples of rural origin, "intercultural" and Afro-Bolivian communities for the provision of tourism services. The law determines that Indigenous communities must participate in planning, organizing and managing any type of CBT (e.g., agri-tourism, ethno-tourism, etc.). Thus, the law stresses that Indigenous communities active participation is what will lead the country's tourism to be sustainable, because they will bring into it their cultural vision of reciprocity, redistribution, complementarity and other values that guided Indigenous life during more than 500 years of colonial exploitation. The Tourism Law positions the thirty-six Indigenous groups that make up the Plurinational State of Bolivia as the country's custodians of the environment and cultural patrimony, asserting that Indigenous peoples are the repositories of knowledge, values, and symbolic processes of reverence to Mother Earth. Indigenous peoples participation is what will lead tourism and tourists to appreciate, respect and preserve the natural and cultural heritage assets of the Plurinational State. As the new political approach toward tourism began to take concrete form, the government began to elaborate a state model for tourism, under which Indigenous people would become the central players (Ministerio de Culturas y Turismo, 2012).

## RECONFIGURING THE INSTITUTIONAL FRAMEWORK FOR TOURISM

Prior to the creation of the Vice Ministry of Tourism in 1997, tourism issues were dealt with by the Tourism Directorate (1937–1972) and later by the Bolivian Tourism Institute (1973–1989). Both institutions were branches of the Ministry of Culture. From 2009 to the beginning of 2010, the Vice Ministry of Tourism became the Vice Ministry of Tourism Industry, a branch of the Ministry of Production Development and Plural Economy. However, at the end of 2010, the Vice Ministry of Tourism became again part of the Ministry of Culture. The Vice Ministry of Tourism has the authority to design and plan national tourism policies and strategies that agree with principles stated by the sustainable national development plan. Accordingly, the Vice Ministry of Tourism designed the country's sustainable tourism model, which is based on the sustainable use of the country's environment and cultural patrimony. In turn, sustainability is based on the Indigenous communitarian forms of living together, respecting intercultural differences and redistributing economic benefits and political power. Hence, the economic and political redistribution implies the participation of each of the territorial administrative government levels (e.g., municipalities) and their civil organizations (see figure 2.1) (Ministerio de Culturas y Turismo, 2015).

The government stressed that the success of the national development plan is contingent on establishing coordination and participation networks among the different levels of state institutions and civil regional and local organizations. To this end, in 2013, the Vice Ministry of Tourism organized a series of meetings with regional and local governments and civil organizations to establish channels of institutional coordination for advancing the implementation of the country's tourism development model. These meetings followed the creation of three new institutional branches: the Tourism Sectorial Coordination Council, the Department Tourism Councils, and the Technical Autonomous Secretariat. The Tourism Sectorial Coordination Council is formed by representatives of the nine departments; they coordinate long-term tourism activities with the Vice Ministry of Tourism. In turn, each of the nine departments has a Department Tourism Council. This council is formed by representatives of local governments (e.g., municipalities), civil urban and rural organizations and Indigenous organizations.[7] According to Law No. 031, "Andrés Ibáñez," Framework for Autonomy and Decentralization,

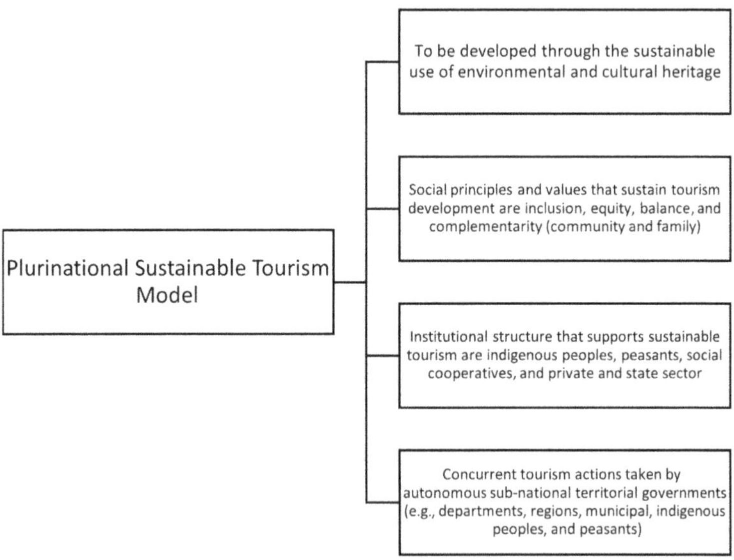


**Figure 2.1 Plurinational Sustainable Tourism Model.** *Source*: Ministerio de Culturas y Turismo, 2015. Elaboration: F. Sonia Arellano-López.

departmental governments have the obligation to coordinate tourism plans and activities with their municipal governments. Although, the Technical Autonomous Secretariat is considered branch of the Tourism Sectorial Coordination Council, it provides technical support to all institutions and organizations involved in tourism planning and implementation (see figure 2.2). This new institutional arrangement suggests the government's commitment to encouraging regional and local governments, civil organizations and particularly, Indigenous organizations participation in the country's tourism development (Estado Plurinacional de Bolivia, 2016:117; Ministerio de Culturas y Turismo, 2015:15–19).

The establishment of the new organizational structure was followed-up by the draft of the country's National Tourism Plan (*Plan Nacional de Turismo—PLANTUR*). The National Plan maintains that tourism is a vital contributor to job and wealth creation, environmental protection, cultural preservation and poverty alleviation. Tourism is an economic and social development strategy that should promote and support the governments'

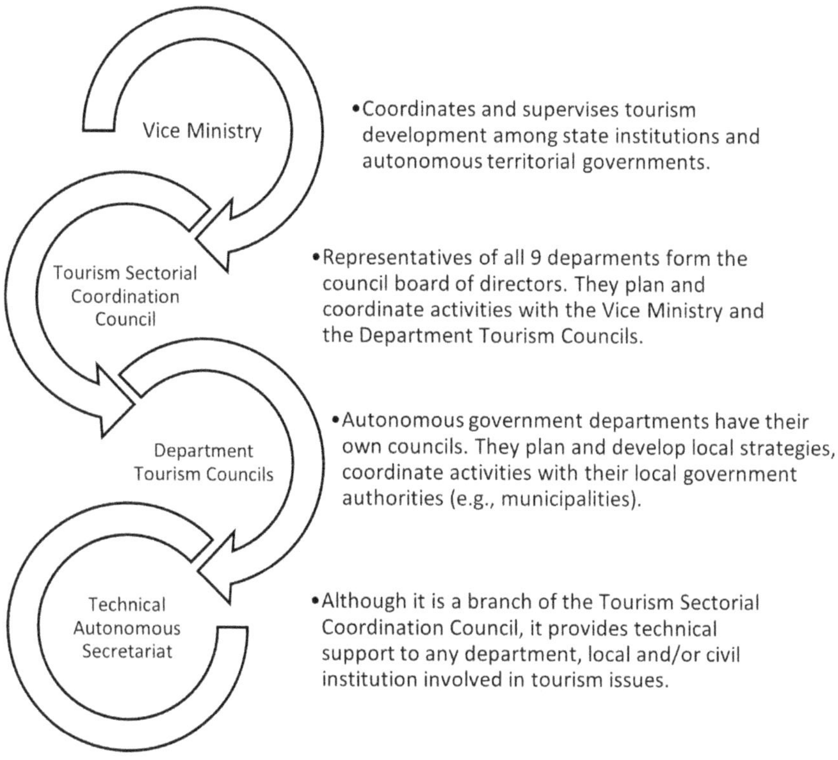


**Figure 2.2 Institutional Coordination Levels.** *Source*: Ministerio de Culturas y Turismo, 2015. Elaboration: F. Sonia Arellano-López.

political and economic principles outlined in the sustainable development plan. It highlights how the country's geographical environmental and Indigenous cultural diversity are hallmarks for promoting tourism. The Plan, similar to the Tourism Law, does not obliterate or hinder private tourism initiatives, but it prioritizes community-based tourism and ecotourism development. The plan emphasizes that these new types of tourism will help reduce the socioeconomic inequality gap that exist in the country; and, put Indigenous peoples to the forefront of the country's sustainable development. Indigenous people will become principal advocates for valuing and preserving their cultural identity and biodiversity in their own territories (Ministerio de Culturas y Turismo, 2014a, 2015).

As the development of community-based tourism and ecotourism became the main component of the country's tourism development plan, the Vice Ministry secured financial support from the National Development Fund System, to design the Sustainable Territorial Management of Tourism Destinations and the Advertisement and Marketing of Tourism Offerings in National and International Markets programs. The Sustainable Territorial Management of Tourism Destinations program's goal was to map, identify and produce an inventory of viable tourism sites for assessing the economic feasibility for future CBT and/or ecotourism initiatives development. The Advertisement and Marketing of Tourism Offerings in National and International Markets program's goal was to prioritize what new types of initiatives could CBT or/and ecotourism offer. Examples include "extreme adventure tourism," which suggests the marketing of the country's environmental diversity; "authentic culture" indicates the marketing of the thirty-six Indigenous groups that make up the Plurinational State of Bolivia and "cultural nature" implies the marketing of the national parks and protected areas. In summary, with the new institutional coordination structure, the draft of national tourism development plan, the tourism inventory sites and the tourism marketing strategies programs, the government had successfully incorporated tourism as vital political and economic component of its national sustainable development plan (Ministerio de Culturas y Turismo, 2015:172–176).

## COMMUNITY-BASED TOURISM AS THE COUNTRY'S SUSTAINABLE DEVELOPMENT FRONTRUNNER

Before to the enactment of Tourism Law and the institutional reorganization, the government set up the legal and institutional basis for launching CBT. During 2006, the National Community-based Tourism Committee was created and the Community-based Tourism Specific Regulations were drafted.

During the same period, the Vice Ministry implemented the Tourism Destinations Sustainable Territorial Management program. This program overlapped with the division of the country into territorial productive development complexes that the government had defined as part of the country's economic program (Ministerio de Culturas y Turismo, 2016: 53–54). Thus, based on biophysical, socioeconomic and/or cultural characteristics that could produce tourist attractions of different categories and hierarchies, the country was divided into five Touristic Macro Regions: Madidi-Amboró, the Precambrian Shield, the Andean, the Chaco, and the Amazon. The macro regions allowed the Vice Ministry to identify tourism attraction sites based on their social and economic characteristics. For example, the Chaco macro region characterized by dry forest, offer potential tourism attractions like the Kaa-Iya National Park, cultural practices of the Guaraní people and the route that Ché Guevara took during his guerrilla campaign of 1966–1967 (Ministerio de Culturas y Turismo, 2012:46–50).

The Tourism Destinations Sustainable Territorial Management program was followed by the design and implementation of five CBT models. The models' sought to calculate an explicitly economic value for cultural heritage and environmental functions of the selected tourism sites and formally integrate these values into the sustainable tourism development plan. They also attempted to show the communities that through this new type of tourism they had an economic alternative that could help them increase their income. These goals were incorporated into the political discourse of Indigenous communities managing their own environmental and cultural resources (Cox Aranibar 2009:124–126; Ministerio de Culturas y Turismo 2014a, b; Viceministerio de Turismo, 2015:12–14).

Each CBT model was designed to offer a particular tourism attraction. The model implemented in Madidi National Park and Natural Area of Integrated Management, located in the upper Amazon river basin of Bolivia, home of the *Takana*, *Lecos* and *Quechua* Indigenous communities, offered the attraction of "nature adventure tourism." The Lake Titicaca model, located in the Andean highlands at 12,500 feet above sea level and considered the world highest navigable lake, offered the attraction of "cultural tourism." It involved the participation of *Aymara* Indigenous communities settled at the lake shores and islands. The salt flats of *Uyuni* and the *Eduardo Avaroa* Natural Reserve, located in the Andean highlands at 13,000 to over 14,000 feet above sea level, offered the attraction of "ecotourism." The implementation of this model was predicated on participation of Aymara-speaking Indigenous communities. The *Tocaña* Afro-Bolivian communities of the *Yungas* area, a narrow band of forest along the eastern slope of the Andes Mountains, offered "adventure tourism." The model implemented in the *Sajama* National Park, located within the Central Andean dry puna ecoregion, with elevations ranging from 13,780 to

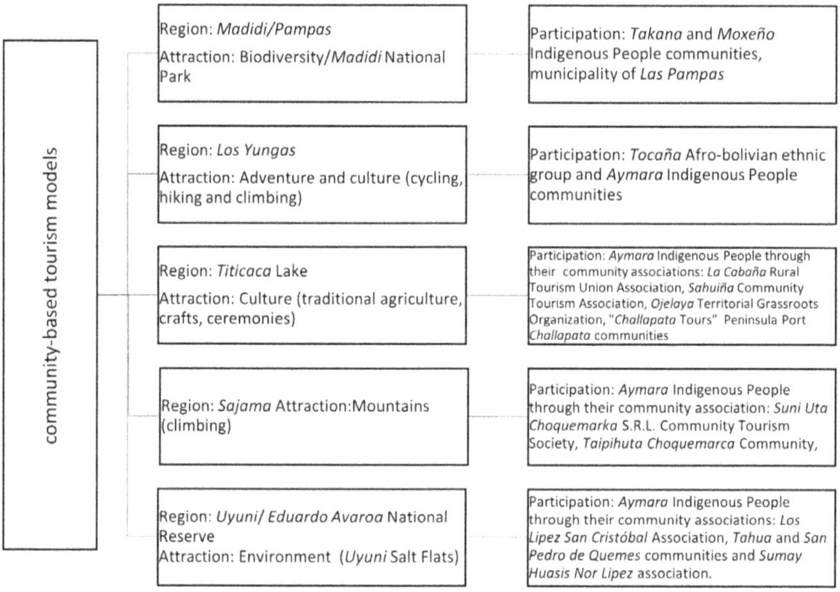


**Figure 2.3 Community-Based Tourism Models.** *Sources*: Viceministerio de Turismo 2014 and 2015. Elaboration: F. Sonia Arellano-López.

21,463 feet above sea level, offered "mountain tourism" (e.g., mountain climbing). *Aymara* Indigenous communities were also involved in this model (see figure 2.3). These models were widely publicized by the Advertisement and Marketing of Tourism Offerings in National and International Markets program through brochures such as "Bolivia: Communities in Tourism Action," "Bolivia: Tourism in Campaign" and "Catalog of Community-based Tourism Initiatives" (Ministerio de Culturas y Turismo, 2015:148–150).

By 2008, approximately eighty-two CBT initiatives had started. By 2011, the Vice Ministry of Tourism reported that the number of initiatives had reached approximately 100, and it confirmed that by 2015 the number had doubled to 200 initiatives. However, the Vice Ministry recognized that a detailed registry and analysis of the number of initiatives reported was necessary. Reports and analyses done by independent consultants showed that not all of the eighty-two community-based initiatives reported in 2008 had the same level of progress. Some were in process of obtaining their working permits and others were building appropriate tourism infrastructure (see table 2.1). The reports found that two types of tourism management characterized these community-based initiatives: the community association and the family association. In the community association, families belonging to Indigenous communities located near or in the tourism site decided to form an association to offer tourism services (e.g., lodging, gastronomy, interpreter, craft trade, etc.). Employment rotated among the community association members, and

**Table 2.1 Community-Based Tourism Initiatives by Department in 2008**

| *Department* | *Total Number CBT* | *CBT Working* | *CBT in Formation* | *CBT & Private** | *Indigenous Communities*** |
|---|---|---|---|---|---|
| **La Paz** | 22 | 14 | 6 | 2 | *Quechua/Aymara/ Tacana/ Moxeño* |
| **Oruro** | 9 | 5 | 4 | - | *Aymara* |
| **Santa Cruz** | 22 | 11 | 6 | 5 | *Ayoreo/Chiquitano* |
| **Potosí** | 10 | 6 | 3 | 1 | *Quechua* |
| **Chuquisaca** | 4 | 4 | - | - | *Quechua* |
| **Tarija** | 4 | 3 | 1 | - | *Quechua* |
| **Beni** | 4 | 3 | 1 | - | *Tsimane/Moseten/ Quechua/ Aymara* |
| **Cochabamba** | 7 | 4 | 3 | - | *Quechua/Trinitario* |
| **Total** | **82** | **50** | **24** | **8** | |

*Sources*: Bustillos, 2008; Cox Aranibar, 2009:125–27.
Elaboration: F. Sonia Arellano-López.
* In these cases the CBT probably has an agreement or has rented the site to a private company.
** The identification of Indigenous communities does not exclude the participation of other Indigenous communities in the initiative (e.g., interculturales or Afro-Bolivians in the case of La Paz).

a percentage of the profits went to a community fund for future community or tourism investment. In the family association, tourism services were carried out by a family in its own home or plot. The family offered rooms for lodging, restaurant, crafts exhibitions, museums, or other activities. The community established a system for balancing the number of services that each participant family would provide and decide the amount of a fixed percentage of earnings to be deposit in the community fund. The profits of the initiative went to the participant family. There were a few cases where private tourism companies and community-based initiatives had agreements. In these cases private tourism companies provided technical assistance (e.g., management training) to the communities or families participating in the initiative. In other cases the community association or family association leased the tourism site for its functioning as in the case of Indigenous communities of *Lica-Tahua* municipalities in the *Potosí* department. In this case the communities formed a commonwealth and hired a private enterprise to manage their touristic site (Ministerio de Culturas y Turismo, 2011:34).

These reports noted that CBT activities did not replaced Indigenous communities' primary production activities nor improved their working conditions. Instead, tourism activities represented economic alternatives that could provide an additional income stream that could be helpful in a time of crisis. The reports also noticed that the stability on the provision of tourism services varied depending on the CBT initiative location. Initiatives working in protected areas and buffer zones provided stable tourism services, while the opposite was true in regions characterized by small and medium agriculture and cattle farms with Indigenous archaeological sites, cultural ancestral agriculture practices, and ceremonial practices.

The reported instability on the provision of services could be associated with the institutional coordination and management of CBT initiatives in the protected areas and buffer zones. For example, in the protected areas and buffer zones the National Protected Areas Service (*Servicio Nacional de Areas Protegidas*—SERNAP), a decentralized government institution, is responsible for designing, implementing, regulating, and controlling tourism development, including CBT activities. CBT activities are coordinated and supervised through the Management Committees (*Comités de Gestión*). These committees are venues where local communities, local and regional governments and Indigenous organizations coordinate decisions on how they will manage their respective protected areas or/and buffer zones. So, when CBT began to sprout the Management Committees became responsible for institutional coordination, as well as for solving disputes among families, communities and/or other third parties involved in CBT activities. Since the National Protected Areas Service oversees these CBT initiatives, they have been characterized as community-based initiatives dedicated to ecotourism (Ministerio de Medio Ambiente y Agua 2011:30).

Protected areas and buffer zones are inhabited by Indigenous people, peasants and intercultural communities. More than ninety municipalities and fourteen Indigenous territories (*Territorio Indígena Originario Campesino, or* TIOC) overlap with protected areas and buffer zones. The National Protected Areas Service estimates that 150 micro, small, and medium tourism services companies operate in the protected areas. In addition to the already established tourism activity, by 2008, the National Protected Areas Service registered thirty-four CBT initiatives functioning in eleven protected areas and buffer zones (see table 2.2). Of those initiatives the National Protected Areas Services informed that the most successful were the *Chalalan* Ecolodge located in the *Madidi* National Park, the *Tomarapi* ecolodge in Sajama National Park, the ecotourism circuit *Lauca* River, *San Miguel del Bala* and the *Kawsay Wasi* an Indigenous tour agency formed by seven Indigenous communities in Carrasco National Park. Certainly, the Vice Ministry of Tourism and the National Protected Areas Services recognized that these successes received a considerable financial and technical support from international organizations, such as the Inter-American Development Bank (IDB), Conservation International (CI), German Technical Cooperation Agency (GTZ), German Development Bank (KfW), United Nations Development Program (UNDP), World Bank (WB), the International Labor Organization (ILO) and other international and national non-government organizations (Ministerio de Medio Ambiente y Agua 2011: 9, 21, 24, 30; Peaty & Portillo, 2009).

**Table 2.2 Community-Based Tourism in National Protected Areas by 2008**

| *Department* | *No. CBT* | *Location* | *Type of Initiative* | *Activities* |
|---|---|---|---|---|
| La *Paz* | 4 | *Madidi* National Park & Natural Area of Integrated Management | Community associations | Eco and cultural tourism |
| | 1 | *Cotapata* National Park & Natural Area of Integrated Management | n.d. | Eco and cultural tourism |
| | 3 | *Apolobamba* Natural Area of Integrated Management | | |
| *Beni* | 2 | *Pilón Lajas* Biological Reserve and TIOC | Community associations | Eco and cultural tourism, archeology |
| *Cochabamba* | 2 | *Carrasco* National Park | Family associations | Eco tourism |
| *Santa Cruz* | 11 | *Amboró* National Park & Natural Area of Integrated Management | Family associations | Eco tourism |
| | 1 | *Noel Kempff Mercado* National Park | n.d. | Eco tourism |
| *Oruro* | 3 | *Sajama* National Park | Community Association *Tata Sajama* | Eco and cultural tourism |
| *Potosí* | 2 | *Eduardo Avaroa* Andean Fauna National Reserve | Family Associations | Eco tourism |
| Tarija | 4 | *Sama* Mountain Range Biological Reserve | In process | Eco tourism and archeology |
| | 1 | *Tariquía* Flora and Fauna National Reserve | n.d. | Eco tourism |
| Total | **34** | | | |

*Sources*: Ministerio de Medio Ambiente y Agua, 2011; SERNAP, 2013; Ríos Araníbar, 2017.
Elaboration: F. Sonia Arellano-López.

## DISCUSSION: THE POLITICAL ADVANTAGES OF SUSTAINABLE TOURISM

The new political Constitution and the government stress that the country's democracy is founded on the Indigenous peoples participation, representation, and ancestral community governance. The government claims that to stop the western capitalist exploitation is to build a plurinational society based on "communitarian socialism." Understanding "communitarian socialism" as regaining Indigenous ancestral values: complementarity, reciprocity, and environmental responsibility. This political assertion, on the one hand, puts the thirty-six Indigenous groups as the vanguard of the country's development. On the other hand, it acknowledges that the country's poverty is centered in the rural areas; 39 percent of the Indigenous population in rural areas lives in extreme poverty (Galindo, 2018:52–56; Ministerio de Trabajo, Empleo y Previsión Social, 2015). So, sustainable tourism provides the government with a political and economic strategy to eradicate poverty and social exclusion. As Goodwin argued (2007:84) after all, poverty is multi-dimensional; it is not only about a lack of monetary resources, but it is also the dimension of exclusion, especially when considering the position of Indigenous communities. Undoubtedly, placing sustainable tourism at the heart of the government's political and economic strategy was not a spur-of-the-moment decision; it has strong political and economic implications for its agenda to promote development in rural areas where most of the people are Indigenous.

Supported by the Constitution and several laws, the government was able to incorporate tourism officially as a development policy, which was followed by strengthening the Vice Ministry of Tourism as the main institution that speaks radically, convincingly, and frequently about the protection of the country's cultural and environmental treasures. The Vice Ministry developed a particular sustainable tourism model that maximizes the potential of tourism for eradicating poverty by developing networks of coordination and participation among different government institutional levels, civil institutions, and principally Indigenous organizations and their communities. The government development plan and tourism model emphatically claimed that Indigenous peoples were not anymore objects of tourism; they were the stakeholders of tourism. They will plan, manage, control, and audit the use of their own environmental and cultural resources.

So, in the pursuit of expanding the sustainable tourism model, the new types of tourism considered to be of low environmental impact and high social and economic inclusion were useful tools for ensuring Indigenous participation, cultural and environmental conservation. They sponsored workshops, seminars, and other activities to push CBT as the banner of the country's sustainable tourism development. CBT offered an alternative to gain something that was lost when the country began a process of development from an agrarian community structure to an industrialized society. The notion of community brought back Indigenous cultural notions of collective decision and action, premises that are opposite to self-interest or individualism that operate in industrialized societies (Harwood, 2010:1910). Guided by this notion of community, the government began to implement its CBT program.

The objective of the implementation of CBT was twofold. First, the government sought to calculate an explicit economic value for cultural heritage and environmental functions of the tourism sites and formally integrate these values into the sustainable tourism development plan. The premise behind the environmental and cultural commodification was that the tourism industry would provide additional economic benefits to the current unfolding sustainable development strategies that the government is implementing. Second, the economic opportunities and benefits derived from CBT will be distributed through ancestral communities' practices of reciprocity, complementarity, and responsibility. Certainly, this type of sustainable CBT planning offered the image of cultural diversity but homogenous Indigenous communities structure. Issues that proved to be complex once CBT initiatives began to sprout all over the country, as, for example, the institutional planning of initiatives in the protected areas and buffer zones. An additional issue that needs to be highlighted is how CBT turned out to be a strategic tool for negotiating complicated political relationships between potentially conflicting groups and the government too. For example, the enforcement of CBT initiatives in the *Isiboro Sécure* National Park, home of the *Mojeño, Yuracaré*, and *Chimán* Indigenous people, provided a temporary resolution to land conflicts between Indigenous people and the government (SERNAP, 2018).

## NOTES

1. This chapter was presented as a paper in the section Engaging Change through Advocacy and Social Action in Heritage Practice in the 79th Annual Meeting of the Society for Applied Anthropology, Portland, Oregon, March 19–23, 2019. Since the SfAA meeting, the Bolivian socioeconomic and political situation has dramatically changed. Evo Morales, the president at the time, was forced from office because he appeared to be engaging in election fraud. He was replaced by a conservative government and was in exile for a couple of years. In 2020, after national elections, Luis Arce a member of Morales' Movimiento al Socialismo (MAS) party won. Mr. Morales returned to the country, but as a consequence of political disagreements with the new leadership of his former party; he currently represents an opposition faction of the MAS. Parallel to this political turmoil, the COVID-19 pandemic halted tourism, with an especially heavy impact on CBT. As a result, the national political and economic discussion has changed, and the situation described in the chapter is very different today. I wish to thank to Michael Painter and Robert Hitchcock who kindly commented on earlier versions of this chapter.

2. See J. Qian, H. Shen, & R. Law, 2018.

3. The WTO (2018:14) reported that during 2012 more than one million international cross-border movements were made by people classified as tourist. By 2017 international arrivals grew to approximately 1.322 billion. By the same year tourism contributed 10.2 percent to the world's GDP, it represented 30 percent of exports in services globally and one out of every ten jobs worldwide is provided by tourism. Tourism is considered the third highest world category in export earnings.

4. See WTO: World Travel and Tourism Council (UK); Earth Council (Costa Rica), 1997.

5. The 2009 Constitution elevates the ILO Convention No. 169 and the UN Declaration of the Rights of Indigenous Peoples the status of national law, the Republican Laws No. 1257 and No. 3760 (ILO, 2009; Ministerio de Autonomía, 2010).

6. The National Institute of Statistics defined Indigenous people only on the basis of the language spoken. Comparing the 2001 census and 2012 census results, the Indigenous population has diminished by 21 percent. However, Indigenous organizations and the general public see *belonging to an Indigenous* people as the more valid criterion. Based on this, the Indigenous population constitutes 62 percent of the total population above fifteen years of age (ILO, 2009:21).

7. The country is divided into nine departments. The departments are divided into cantons and the cantons into municipalities. There are approximately 327 municipalities in the country.

## REFERENCES

Arce Catacora, L. A. (2011). *El Nuevo Modelo Económico, Social, Comunitario y Productivo.* La Paz, Bolivia: Ministerio de Economía y Finanzas Públicas.

Bustillos, M. J. (2008). *Informe sobre emprendimientos turísticos comunitarios en Bolivia.* Retrieved March 2018, from Redturs Web site: http://www.redturs.org

Cox Aranibar, R. (2009). *Turismo Indígena y Comunitario en Bolivia.* La Paz, Bolivia: Plural Editorial.

Duffy, R. (2002). *A trip too far: ecotourism, politics and exploitation.* Earthscan.

Estado Plurinacional de Bolivia. (2016). *Plan de Desarrollo Económico y Social 2016-2020, En el Marco del Desarrollo Integral para Vivir Bien.* La Paz - Bolivia: Estado Plurinacional de Bolivia.

Fundación Tierra. (2011). *Territorios Indígena Originario Campesinos en Bolivia. Entre la Loma Santa y la Pachamama.* Fundación Tierra. La Paz-Bolivia: Scorpion 2460067.

Galindo, M. (2018). *Descriptores del Estado Plurinacional de Bolivia (2009-2017).* La Paz, Bolivia: Centro Boliviano de Estudios Multidisciplinarios (CEBEM).

Goodwin, H. (2007). Indigenous tourism and poverty reduction. In: R. Butler, & T. Hinch (eds.), *Tourism and Indigenous Peoples: Issues and Implications* (pp. 84–94). Great Britain: Elsevier Ltd.

Hall, M. C., Gossling, S., & Scott, D. (2015). The evolution of sustainable development and sustainable tourism. In: M. C. Hall, S. Gossling, & D. Scott (eds.), *The Routledge Handbook of Tourism and Sustainability* (pp. 2–10). Routledge.

Instituto Nacional de Estadística (INE). (2015). *Censo de Población y Vivienda 2012, Bolivia: características de la población.* La Paz-Bolivia: Instituto Nacional de Estadística (INE).

International Labor Organization (ILO). (2009). *Indigenous & Tribal People's Rights in Practice: a guide to ILO Convention 169.* International Labor Organization (ILO). Geneva: ILO Publications.

Ministerio de Autonomía. (2010). *Los Pueblos Indígena Originario Campesinos en el Nuevo Modelo del Estado Plurinacional con Autonomías.* La Paz-Bolivia: Ministerio de Autonomía.

Ministerio de Culturas y Turismo. (2000). *Bolivia te espera.* La Paz-Bolivia: Estado Plurinacional de Bolivia.

Ministerio de Culturas y Turismo. (2011). *Plan Nacional de Turismo 2012-2016.* La Paz-Bolivia: Ministerio de Culturas y Turismo.

Ministerio de Culturas y Turismo. (2012). *Ley General de Turismo "Bolivia te Espera".* La Paz-Bolivia: Ministerio de Culturas y Turismo.

Ministerio de Culturas y Turismo. (2014a). *Bolivia: Turismo en Marcha.* La Paz-Bolivia: Ministerio de Culturas y Turismo.

Ministerio de Culturas y Turismo. (2014b). *Guía Turística. Emprendimientos Comunitarios.* La Paz-Bolivia: Ministerio de Culturas.

Ministerio de Culturas y Turismo. (2015). *Plan Nacional de Turismo 2015-2020-Agenda Turistica PLANTUR 2025.* La Paz-Bolivia: Ministerio de Culturas y Turismo.

Ministerio de Culturas y Turismo. (2016). *Plan Estratégico Institucional (PEI) 2016-2020.* La Paz-Bolivia: Fondo de Fomento a la Educación Patriótica.

Ministerio de Medio Ambiente y Agua. (2011). *Plan Estratégico para el Desarrollo del Turismo en el Sistema Nacional de Áreas Protegidas. Construyendo los Destinos Turísticos Más Auténticos de Bolivia.* La Paz-Bolivia: Ministerio de Media Ambiente y Agua.

Ministerio de Trabajo, Empleo y Previsión Social. (2015, Septiembre). *Biblioteca Enlace.* Retrieved October 18, 2018, from http://www.enlace.comunicacion.gob.bo

Mowforth, M., & Munt, I. (2009). *Tourism and Sustainability: Development, Globalization and New Tourism in the Third World* (3rd edition). New York: Routledge.

Peaty, D., & Portillo, A. (2009, March). *Community-based tourism in Bolivia: Projects and perspectives.* Retrieved February 2018, from http://www.ritsumei.ac.jp/acd/cg/ir/college/bolletin/Vol.21-3/06Peaty_Portillo.pd

Postero, N. (2017). *The Indigenous State. Race, Politics, and Performance in Plurinational Bolivia.* Oakland, CA: University of California Press.

Qian, J., Shen, H., & Law, R. (2018, July 31). *Research in Sustainable tourism: A Longitudinal Study of Articles between 2008-2017.* Retrieved July 31, 2018, from Sustainability : http://www.mdpi.com/jounal/sustainability

Ríos Araníbar, G. (2017). *Turismo de base comunitaria en Bolivia: nuevo modelo de gestión local para el desarrollo socio-económico e intercultural de las comunidades índigenas.* Retrieved March 6, 2018, from http://www.admin.egal2017 .bo

Servicio Nacional de Areas Protegidas. (2013). *Guía Turística de las Areas Protegidas de Bolivia.* La Paz-Bolivia: Viceministerio de Turismo / SENAP.

Servicio Nacional de Areas Protegidas. (2018, November 13). Retrieved from SENAP: http://www.sernap.go.bo

Sturtevant, Chuck (2023) "The settler roots of Plurinational Bolivia: State-sponsored indigenous colonization on Bolivia's Amazonian 'frontier." *Settler Colonial Studies* 13(3): 419–437.

UDAPE Bolivia. (2015, September). *UDAPE Bolivia.* Retrieved September 4, 218, from UDAPE Bolivia: http://www. gob.bo

Viceministerio de Turismo. (2015). *Bolivia: Comunidades en Acción Turística. Catálogo Promocional PNTC Componente II. Apoyo directo a los Emprendimientos Turísticos Comunitaríos (ETC).* La Paz/Bolivia: Ministerio de Culturas y Turismo.

Wearing, S., & Neil, J. (2014). *Ecotourism: Impacts, Potentials and Possibilities* (2nd edition). New York: Routledge.

World Tourism Organization (UNWTO). (2018). *Tourism for Development - Volume I: Key Areas for Action.* Madrid/Spain: World Tourism Organization (UNWTO).

WTO: World Travel and Tourism Council (UK); Earth Council (Costa Rica). (n.d.). *Agenda 21 for the travel and tourism industry: Towards environmentally sustainable development.* 1997: WTO.

*Chapter 3*

# Challenges of Ecotourism among the San of Botswana, Namibia, and Zimbabwe

Robert K. Hitchcock, Melinda C. Kelly, and Wayne A. Babchuk

Issues relating to Indigenous and minority peoples in the Kalahari Desert region of southern Africa center upon the impacts of tourism and whether tourism represents a source of sustainable economic development for Indigenous peoples at the local community level (Hitchcock 1997; ‡Oma and Thoma 1998; Mbaiwa 2008, 2012, 2017; Koot 2013, 2018; Green and Saarinen 2022). The World Bank (2018, 2020a, b) sees wildlife and nature tourism as a key means of supporting sustainable livelihoods. The United Nations World Tourism Organization, the European Union, and various government development agencies (e.g., the Swedish International Development Agency, SIDA) promote ecotourism through providing assistance to development-related ecotourism projects (Fennell 2020). Currently, the government of Botswana and the United Nations Development Program have a large-scale project in western Botswana which is devoted in part to ecotourism and community livelihoods development (Government of Botswana and United Nations Development Program 2021).

Tourists, for our purposes here, are defined as those people who travel away from home for pleasure and stay at least one night and less than a year (Crick 1984; World Tourism Organization 1993; Mtapuri and Giampiccoli 2018). We differentiate between recreational tourists and businesspeople who undertake tourist visits during the course of their work in a country. Host populations who deal with tourists often see tourism as a double-edged sword, bringing both costs and benefits (Chambers 1997; Stonich 1998; De Kadt 1979; Andereck 2005; Brondo 2013). This chapter provides a critical analysis of ecotourism practices in Botswana, Namibia, and Zimbabwe.

## TOURISM POLICIES AND STRATEGIES

Policies involving tourism have evolved significantly in Botswana, Namibia, and Zimbabwe over the past several decades (Republic of Botswana 1990, 1991, 2009a, b, Republic of Namibia 2000, 2008; Republic of Zimbabwe 2016). Tourist numbers in the three countries increased substantially in the past decade, at least until the coronavirus pandemic was declared beginning in March 2020. Tourism contributed 13.1 percent of Botswana's economy in 2019, up from 6.3 percent in 2000. In the case of Namibia, tourism contributed 10.9 percent of Gross Domestic Product (GDP) in 2019. Travel and tourism contributed 6.6 percent to Zimbabwe's GDP in 2019. In all three cases, there were fluctuations in the percentages of tourism contributions to the economies over time.

Community-based tourism (CBT), while mentioned in the tourism-related policy documents of the governments of Botswana, Namibia, and Zimbabwe, does not appear to get the emphasis that nature-based tourism does in these three countries, although it represents a sizable proportion of the tourism industry. Archaeotourism, which includes tourist visits to archaeological sites (see Giraudo 2016 and Porter 2020), is particularly important in Zimbabwe, home of the Great Zimbabwe World Heritage Site and Matobos Hills National Monument, but it is also important in Botswana, as seen at the Tsodilo Hills World Heritage Site, and in Namibia, at the Twyfelfontein World Heritage Site and the Brandberg World Heritage Site.

Many of the tourists who visit southern Africa can be characterized as ecotourists or adventure tourists, those who seek unique experiences in other places can be said to fall in the category of leisure tourists. There are also tourists who can be considered business tourists, those individuals coming into a country for the purpose of conducting business or taking part in conferences. A sizable number of these business tourists occasionally visit protected areas or places of cultural and natural significance, particularly those on the outskirts of capital cities.

Specialized tourists include those who visit Africa to engage in trophy, recreational, or safari hunting (see Lindsey, Roulet, and Romañach 2007; Crosmary, Cote, and Fritz 2015; Naidoo et al. 2016; Angula et al. 2018). In some cases, as in Nyae Nyae and N≠a Jaqna, Namibia, safari hunting operators and their clients work with Indigenous guides who accompany them on hunts. The safari groups generate employment and salaries for hunting guides, and they generally provide the communities with meat from the hunts and, in some cases, medical supplies. Joint venture agreements existed between the communities and the safari hunting companies which outline the responsibilities of both companies and the communities. In the case of Botswana, however, many of these arrangements have been transformed since the 2014 hunting ban and the COVID-19 pandemic of 2020–2021, with a significant reduction in benefits to the communities.

Cultural tourism, sometimes called ethno-tourism, can be defined as tourism that is concerned with a society or region's culture, especially the lifestyles, traditions, values, rituals, religion, art, ideas, and behaviors of that society (Guenther 2002). One type of cultural tourism is heritage tourism, which enables visitors to see places that are significant from a cultural, archaeological, historical, architectural, and/or religious standpoint. It is estimated that cultural tourism and heritage tourism draw about 10–15 percent of international tourists who visit southern Africa (Haratsebe, Moswete, and Saarinen 2016). A significant percentage of these tourists come from North America, Europe, Asia, and from the Southern African Development Community.

Some of the tourists who visit local communities are free, independent travelers (Crick 1984) or, as some people call them in southern Africa, "hippie tourists." These kinds of tourists, according to local communities, tend to be more interactive with local people than those tourists who come to southern Africa to see wildlife. Interviews of local communities reveal that there are mixed perspectives on "free independent travelers." Some community members say that they like these kinds of tourists because they generally show intense interest in local communities. Conversely, some community members say that hippie tourists do not generally bring much money with them, and they sometimes stay for a long time in the villages and end up depending on local people for food, water, and other commodities.

## ECOTOURISM IN THE KALAHARI REGION

For purposes of the present study, we will concentrate on several areas in the Kalahari where tourism has had significant effects: including the Nyae Nyae region of Namibia, the Dobe-/Xai/Xai region, the Tsodilo Hills region, Ghanzi District of Botswana, and the western part of Zimbabwe south of Hwange National Park (see figure 3.1).

All these areas have attracted tourists for decades, in part because of the people who reside there. Perhaps the best known of the San whom tourists visit are the Ju/'hoansi (formerly called the !Kung) who reside in the northeastern part of Namibia and the northwestern part of Botswana. The Ju/'hoansi have been the people who have had the most tourists visit them over the past sixty years (Marshall 1976; Lee 1979, 2013; Biesele and Hitchcock 2013; Hays and Hitchcock 2020). The northern Kalahari where the Ju/'hoansi reside is among the best-known areas in the world where local

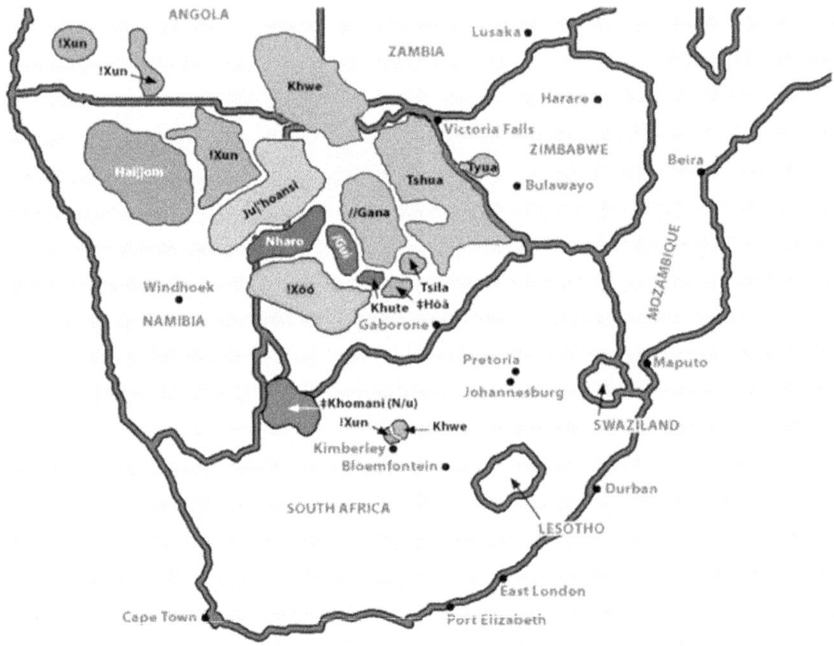


**Figure 3.1 Map of the Distribution of San Peoples in Southern Africa.** *Source*: Kalahari Peoples Fund (KPF).

**Table 3.1 Numbers of San Botswana, Namibia, and Zimbabwe**

| *Country* | *Population Size (2022)* | *Size of country (in km²)* | *Numbers of San (National)* |
|---|---|---|---|
| Botswana | 2,384,246 | 581,730 | 68,500 |
| Namibia | 2,727,409 | 824,292 | 38,000 |
| Zimbabwe | 15,121,004 | 390,757 | 2,800 |
| Totals | 20,232,659 | 1,796,779 | Approx 109,300 |

*Source*: Data obtained from the Southern African Development Community; The World Factbook (2022); Ethnologue (www.ethnologue.com), accessed October 3, 2023, and from fieldwork.

people derive a significant portion of their livelihoods from hunting and gathering. They are thus of great interest to tourists, journalists, filmmakers, researchers, and development workers (see, for example, van der Post 1958; Marshall 2003; Tomaselli 2021). Table 3.1 shows the number of San in Namibia, Botswana, and Zimbabwe.

An issue that seriously affects the well-being of Ju/'hoan communities in Nyae Nyae and Nhoma in Namibia is the establishment of ecotourism operations, which sometimes remove well-known healers, trance dancers, trackers, and craft producers from their communities and transport them far from their homes to interact with and entertain tourists. One example of such activities is that of Naankuse, a tourism operation near the Windhoek Airport in Central Namibia, which brought from their northern village the prominent late healer /Kunta Boo and his wife !Nae, among other Ju/'hoansi.

When tourists were asked about why they visit Nyae Nyae, Dobe, /Xai/Xai, and Tsodilo, they frequently said that they have been inspired by films focusing on the Ju/'hoansi, particularly those of Laurens van der Post and John Marshall (see, for example, van der Post 1958; Marshall 2003). There are also films that depict in detail the interactions between the Ju/'hoansi and tourists (see Zips and Zips-Mairitsch 2015; Stadler 2016). Some of the tourists who come to this part of southern Africa say that they do so in hopes of seeing both wildlife and people. This is true, for example, of the Central Kalahari Game Reserve in Botswana, made famous by National Geographic and British Broadcasting films and by books such as *Cry of the Kalahari* by Delia and Mark Owens (Owens and Owens 1984). Sadly, it was on the recommendations of conservationists that the San and Bakgalagadi were removed from the Central Kalahari in 1997 and 2002, in part to make way for tourism (Owens and Owens 1981; Sapignoli 2018).

To the north of the Nyae Nyae area is the Khaudum National Park, which was declared in 2007 (prior to that it was a game park). Khaudum is 3,841 $km^2$ in size and is known for its remoteness and its large numbers of elephants. Ju/'hoan San were required to move out of Khaudum National Park in 2007, relocating to Nhoma and Aesvolness in the northern part of the N≠a Jaqna Conservancy. Only one tourist camp in Namibia can be said to be San-owned, which is Treesleeper Camp north of the Etosha National Park (Hüncke and Koot 2012). There are a number of farms south of Etosha National Park which are now in the hands of Hai//om communities, though the tenure arrangements have yet to be finalized.

In Botswana, several San communities have engaged in ecotourism operations, in areas ranging from the Tsodilo Hills to the Makgadikgadi Pans (see figure 3.1). In Ghanzi District, Naro San from D'Kar own Dqae Qare Game Farm, a CBT operation which provides game viewing, dancing, and other activities. Dqae Qare is located approximately 11 km north of Ghanzi and 20 km south of the D'Kar community and is the only San-owned freehold land in Botswana. It was purchased by the Dutch government for the Naro San in 1997. As of 2020, there were twelve employees on the farm along with two farm managers. As many as forty people from D'Kar come to the farm to take part in dances for which they are paid (for a discussion of the dance groups and other activities at Dqae Qare see Bollig et al. 2000). The Dqae Qare San Lodge generated P275,000 (US$24,000) in 2010 and P2.8 million (US$248,000) in 2019. Income to the D'Kar community averaged P500,000 (US$44,250) per year over the past decade. In 2019, Dqae Qare lodge had sixteen rooms and also supported an area for camping for tourists. There were 5,700 international tourists who visited Dqae Qare in 2019. Because of loans sought to support the lodge in the past two years, the board of the lodge owes some US$52,000 to the bank, something they were desperately trying to pay off. This effort was made much more complex with the declaration of COVID-19 in Botswana in March 2020. A crowdfunding effort obtained some of the funds necessary, but by March 2023 it appeared that the bank had foreclosed on the loan and Dqae Qare was no longer open.

An examination of the activities at Dqae Qare and other tourism operations in southern Africa reveals that the San in these places engage in a wide range of cultural heritage activities aimed at tourists, such as: bush walks, demonstrations of plant gathering, demonstrations of hunting skill displays, observations of tracking skills, traditional dancing, trance healing ceremonies, trap making using Sansevieria twine, fire-making, arrow poison beetles collection and manufacture of poison, tool manufacture and use, ostrich eggshell bead manufacture, traditional games, blacksmithing, children's games, and storytelling. Similar activities are also demonstrated in cultural heritage sites in Namibia and South Africa.

Judging from tourist interviews and government reports, some of the tourists who visit the Kalahari do so in order to take part in cultural tourism. Others enjoy visiting "model villages" or "living museums" (Zips 2015; Haratsebe, Moswete, and Saarinen 2016; Ritterband 2018). Some of these living museums and model villages are the product of local community-based efforts. Others, such as the "Living Museums" of the Living Culture Foundation of Namibia in Namibia, combine local community efforts with those of a national-level non-government organization.

Living museums that have national or international connections tend to provide a portion of the funds and employment that they generate to local community members. This can be seen, for example, in the cases of Grashoek in the N≠a Jaqna Conservancy and the "Little Hunters" Museum at //Xao//oba in the Nyae Nyae Conservancy in northeastern Namibia (Zips and Mairitsch-Zips 2015; Hitchcock 2015, 2019a). See figure 3.2 for a photo of an individual visiting Ju/'hoansi in Nyae Nyae. In Namibia, the living museums provide 10 percent of their earnings to the conservancies where they are located in addition to providing payments to community participants in tourism-related activities. Table 3.2 presents data

**Figure 3.2 Tourist Visiting the Region.** *Source*: Photo by the author.

on the living museums and cultural villages that can be found in Namibia, Botswana, and Zimbabwe.

Living museums and cultural villages were doing fairly well until the coronavirus pandemic was declared in March of 2020. Virtually all of them lost money in 2020 and 2021. Dqae Qare and others sought loans to stay afloat, while some were forced to shut down permanently.

Some living museums are based in conservancies, areas of communal land in Namibia where communities have rights over wildlife under government legislation. San peoples are in the majority in two conservancies (Ju/'hoansi in the case of Nyac Nyae, and !Kung, !Xun, Khwe, Vasekele, Hai//om, and Ju/'hoansi in the case of N≠a Jaqna) where they have a long history of dealing with tourists, especially Nyae Nyae (Biesele and Hitchcock 2013; Cole 2018). N≠a Jaqna has had less experience with international agencies and has faced several challenges, including the large-scale immigration of non-San people from other parts of the country who have established cattle posts and erected fences (Hitchcock 2012; Welch 2018). The presence of large numbers of non-San people, cattle, and fences has tended to discourage tourists from visiting the area. The N≠a Jaqna Conservancy, with the support of the Legal Assistance Center in Windhoek, took some thirty-five illegal fencers and grazers to court

**Table 3.2 Living Museums and Cultural Villages in Botswana, Namibia, and Zimbabwe**

| *Name of Living Museum or Cultural Village and Date of Founding* | *Location* | *Population Size* | *Activities Undertaken by Residents and Visitors* |
|---|---|---|---|
| Living Museum of the Ju/'hoansi San, Grashoek, July 2004 | Grashoek, N≠a Jaqna Conservancy, Otjozondjupa Region, Namibia | 250 people, !Kung and Ju/'hoansi San | Dances, gathering trips, guided walks, camping, and crafts manufacture and sales |
| Little Hunter's Museum of the Ju/'hoansi San, March 2010 | //Xa/hoba, Nyae Nyae Conservancy, Otjozondjupa Region, Namibia | 200 people, Ju/'hoansi San | Dances, hunting trips, gathering trips, guided walks, and crafts manufacture and sales |
| Nhoma Safari Camp and Nhoma Community | Nhoma, Nyae Nyae area, Otjozondjupa Region, Namibia | 250 people, Ju/'hoansi San, some !Kung San | Dances, hunting trips, gathering trips, guided walks, and crafts manufacture and sales, including high quality sandals |
| The Living Museum of the Damara, February, 2010 | 8 km from Twyfelfontein, a World Heritage Site, 2007, Namibia | No data, total population of Damara 100,000 | Cultural village, traditional life, bush walks, dancing, and singing |
| The Living Museum of the Mafwe, February 2008 | 22 km north of Kongola in the Zambezi Region, Namibia | Singalawe Village nearby has 21 people working at the Living Museum | Cultural village, traditional life, bush walks, dancing, singing, and crafts |
| The Living Museum of the Mbunza, October 2011 | 14 km west of Rundu, Kavango East Region, Namibia | Samsitu Lake, Kavango Region, 115,447 total in region | Cultural village, traditional life, bush walks, dancing, and crafts |
| The Living Museum of the Ovahimba, November 2016 | 40 km north of Opuwo on C43, road to Epupa Falls | Kunene Region | Cultural village, traditional life, bush walks, camping, and crafts |
| N/a'an ku sê, wildlife sanctuary and cultural village | 30 km west of Windhoek | 3,200 hectare wildlife sanctuary | Cultural village, wildlife viewing, bush walks, dancing, and healing |

| | | | |
|---|---|---|---|
| /Xai/Xai Community, Botswana, October 1997, size 20 $km^2$ | /Xai/Xai, North West District, Botswana | 500 people, Ju/'hoansi San and Mbanderu (Herero) | Dances, hunting trips, gathering trips, guided walks, and crafts manufacture and sales |
| Tsodilo Hills World Heritage Site, 2001, size 225 $km^2$ | Tsodilo Hills, North West District, Botswana | 350 people, 100 Ju/hoansi San, 250 Mbukushu | Dances, guided walks, and crafts manufacture and sales |
| Dqae Qare Game Farm, 1998, size 75 $km^2$ | Ghanzi District, Botswana | 40 people, Naro San | Dances, guided walks, crafts manufacture and sales, and plant identification |
| D'Kar Village, 1986, size 30 $km^2$ | Ghanzi District, Botswana | 1,200 people, Naro San, Bakgalagadi, Herero | Museum, cultural center, art project, and art and crafts manufacture and sales |
| Tsabong Camel Ecotourism Park, 36 $km^2$ | Kgalagadi District, Botswana | Three villages: Tsabong, Mashele, and Maubelo, Tsamama Community Trust | Ecotourism park in communal village area |
| Manxotae, 1975, size 4 $km^2$ | Central District, Botswana | 800 people, Tshwa San, Kalanga, Bamangwato | Involvement in Nata Bird Sanctuary, communal land |
| Dlamini Village, Tsholotsho District | Matabeleland North, Zimbabwe, size 0.5 sq km | 150 Tshwa, Kalanga | Cultural museum |

*Source*: Created by authors.

*Note*: Data obtained from Botswana, Namibia, and Zimbabwe government and from non-government organizations and institutions such as the Living Museums Foundation of Namibia.

in 2013 and won the case in 2015 (van der Wulp 2016). Due to lack of enforcement, most of these illegal immigrants have yet to remove their cattle and fences, despite government orders to do so (van der Wulp and Hebinck 2021).

In Nyae Nyae, pastoralists with their herds entered the conservancy area in April of 2009 (Hays 2009; Hays and Hitchcock 2020; Begbie-Clench 2016). While the Namibia police, with the encouragement of the Ju/'hoansi Traditional Authority and the Nyae Nyae Development Foundation, were initially able to get some of the cattle confiscated, the number of immigrants with herds has increased over time. Part of the problem is that Tsumkwe is a municipality, and the conservancy management committee, the local traditional authority, and the community forest management committee do not have the power to evict the illegal immigrants. A legal case against several of the illegal grazers was filed with the High Court in Namibia in July 2015, but the incomers have sought support from the president of Namibia (March 2018), who potentially could rule in favor of the pastoralists, putting the Nyae Nyae Conservancy and the Ju/'hoansi and their supporters at great risk. Ju/'hoansi in Nyae Nyae are concerned that the presence of sizable numbers of pastoralists with their large herds of cattle will have impacts on the grazing and water resources as well as on tourism.

Several Ju/'hoansi also told us in interviews in 2019 that the cattle are affecting the wildlife distributions and gathering potential. They also suggested that tourists are not excited about seeing cattle but prefer seeing wild animals and people following their traditional lifestyles, who represent a small proportion of contemporary Ju/'ansi. NGOs, tourism companies, and conservancy management committees often argue against raising cattle in these areas because of the environmental impacts and the effects livestock have on Ju/'hoan livelihoods, well-being, and tourism potential.

While statistics are not kept on the number of tourists who visit Tsumkwe and the Nyae Nyae area, it is estimated that approximately 2,000–3,000 visitors a year come to Nyae Nyae and perhaps 300–400 to Nhoma and the N≠a Jaqna Conservancy. According to interviews, Ju/'hoansi, !Xun, and Khwe San maintain that they gained some benefits from the conservation, tourism, and poverty-alleviation efforts of the communal conservancy management committee and the NGOs working there. The Nyae Nyae Conservancy, the oldest communal conservancy in the country, generated over N$5 million (US$417,600) through its activities in 2017, while the N≠a Jaqna Conservancy, the largest communal conservancy in Namibia, generated over N$1 million (US$83,520) (data from the Nyae Nyae Conservancy and the N≠a Jaqna Conservancy Management Committees and the Nyae Nyae Development Foundation of Namibia, 2018).

The social, economic, and environmental impacts of tourism in these areas have been examined by Van Der Burg (2013); Hitchcock (2015); Zips (2015) and Ritterband (2018), among others. There are some tensions between the tourists and the conservancies because, in many cases the tourists who enter the area do not stop at the conservancy offices to pay the required fees. As far as we are aware, the two private tourist facilities in Nyae Nyae and Nhoma did not pay a percentage of their profits to the conservancies, thus reducing the funds available to conservancy members. It appears to be difficult to enforce the conservancy rules. This has changed in the case of Nhoma with the 2021 purchase of Nhoma Safari Lodge by Flora Blommaert and Moritz von Hase, who are community-oriented and have worked out ways to ensure that the Nhoma community benefits from tourism.

People in the two conservancies and the management committees want tourists to follow a set of rules which they have outlined on handouts that they give to tourists when they stop at their offices (Biesele and Hitchcock 2013:209–212; Hitchcock 2015, 2018; Kallie N!aice, Leon Tsamkxao, personal communications, 2017). The San of Nyae Nyae and the N≠a Jaqna Conservancies generally say that they see tourism as an important part of their economies and livelihoods, and they would like to see tourist numbers increase. At the same time, they want to ensure that the tourists act in a socially and environmentally responsible manner. They are particularly concerned with the ways that tourists chase after animals with their vehicles, attempting to photograph them, and the impacts that the tourists have on the landscape, leaving trash behind and depleting the areas of firewood. They also say that they prefer that tourists take photographs of them wearing contemporary clothing rather than asking them to shed their modern clothing and exchange them for skins.

## TOURISM IN SOUTHERN AFRICA

A sizable number of tourists who visit southern Africa say that they travel there specifically to see wildlife and natural wonders such as Etosha National Park or Victoria Falls. Wildlife tourists tend to want to see the "Big 5"—lion, leopard, buffalo, rhinoceros, and elephant. A smaller percentage of tourists aim to visit cultural heritage localities such as the Tsodilo Hills and Drotsky's Caves (Gcwihaba) in Botswana. Nature tourism also attracts people to places such as the Nyae Nyae pans in Namibia. Some of the tourists to whom we spoke wanted to go to Nyae Nyae pans because of the large numbers of water birds that visit the pans. There were also tourists who expressly wanted to see elephants. Our interviews indicated that some of the tourists were pleased to see people as well, though that was not their main reason for visiting Namibia.

Much of the tourism in Botswana, Namibia, and Zimbabwe can be seen as largely an elite activity. Those international tourists who visit southern Africa can pay as much as US$10,000–$20,000 per person for their visits, depending on where they go and what they do. Safari hunters can pay as much as US$50,000–$100,000 for their hunts (Lindsey et al 2007). There is some benefit to local communities from safari hunting, since the companies hire local trackers and domestic workers. But by and large, the safari camps are located far from communities, which limits community benefits.

Those free independent traveler tourists who attempt to camp and to pay as little as possible for tourism activities tend to be looked down upon by both governments and tourism companies, who prefer to have high-end tourists. From the community perspective in southern Africa, the preference is for both low-end and high-end tourists because they provide different benefits. Countries such as Botswana have deliberately set high prices for entrance into protected areas—US$150 per day or more—to increase profits and to reduce the impact of tourist numbers. Namibia has not followed the Botswana model, which the Namibians see as problematic both from the standpoint of increasing tourist numbers and supporting elite tourists to the exclusion of a broad range of tourists.

One way that tourism has been promoted by international agencies such as UNESCO is to declare areas as World Heritage Sites (WHS) (Taylor 2010; 2014; Buurdeau, Gravari-Barbas, and Robinson 2016; Giraudo 2016). In Botswana, two areas were declared WHS: The Tsodilo Hills in 2001 and the Okavango Delta in 2014. The Tsodilo Hills witnessed a substantial increase in tourist visits since it was declared a World Heritage Site (Keitumetse and Nthoi 2009). In 2017, over 20,000 visitors came to Tsodilo, according to the Botswana National Museum staff based at Tsodilo (personal communication, 2017). The number of tourists visiting Tsodilo dropped off substantially in 2020–2021 due to the coronavirus pandemic (Botswana National Museum visitor data 2022).

Research data reveals that in the Tsodilo Hills, the commodification of tourism has led to gender and ethnic differences in access to income, employment, and other benefits (Giraudo 2016; Mbaiwa 2016a). It is mostly males who serve as guides in the hills for tourists, while it is women who make and sell crafts to tourists. In speaking to Ju/'hoansi and Mbukushu who live either adjacent to or some distance away from the Tsodilo Hills, women of both groups point out that they feel that they have been marginalized by the expansion of tourism in Tsodilo. There is competition between the two groups (Ju/'hoansi and Mbukushu) when it comes to engaging in dances and other traditional activities for the tourists. There is a certain degree of ethnic tension in Tsodilo which has arisen in part over competition for the tourists who visit there. Ju/'hoan San and Mbukushu living near the Tsodilo Hills World Heritage Site were told in 2020 that they would have to be relocated to accommodate gas and oil prospecting by ReconAfrica, a Canadian company, a position that was endorsed by the Botswana government (Barbee and Nash 2020; Tsodilo Hills Village Development Committee, personal communications, 2020, 2021, 2022).

There are other land-use conflicts in these areas, one example being the village of /Xai/Xai, which claims rights over Gcwihaba and its caves (Hitchcock et al. 2015). The Gcwihaba Caves, which have been proposed by Botswana as a World Heritage Site, were officially opened in November 2017 by the president of Botswana, Sir Seretse Ian Khama (Keaketswe, 2017). /Xai/Xai residents in the past saw Gcwihaba as their *n!ore* (traditional territory). However, the government plans to allow a private company to build a lodge there to host tourists who want to visit the caves. No consultations have taken place with the nearby /Xai/Xai community about these plans, despite international rules concerning free, prior, and informed consent regarding development issues and Indigenous peoples (Food and Agriculture Organization of the United Nations 2016; United Nations Development Programme 2017). The Gcwihaba Caves not only contain important archaeological materials, but also support several colonies of bats (Commerson's leaf-nosed bats, *Hipposideros commersoni* (Chiroptera: Hipposideridae), African honeybees (*Apis mellifera*), and some unusual insects and other species. Concern has been expressed by tourists and residents of /Xai/Xai about the possibility of bats carrying diseases that could affect humans, a fear which has been exacerbated by the correlations made by scientists about bats as a possible origin of coronavirus (SARS-CoV-2) (for a discussion of bats as viral reservoirs, see Streicker and Gilbert 2020).

At /Xai/Xai, which hosts tourists, either self-driving ones or clients brought in by safari operators such as Uncharted Africa Safaris, local Ju/'hoansi give demonstrations of plant collecting, tracking, fire-making, and dancing. Tourists have sometimes asked to see arrow poison preparation. A few older males know how to do this, but they are reluctant to show tourists because they consider it to be very dangerous, and they feel that the knowledge is secret. Ju/'hoansi and a few Herero at /Xai/Xai sell crafts to tourists, and the /Xai/Xai Dance Troupe performs for tourists in addition to going to the yearly Kuru Dance Festival at Dqare Qare in Ghanzi.

Women say that they are largely excluded from hunting tourism operations other than serving as cooks and camp cleaners, something that is common across Africa when it comes to safari hunting tourism (Hunter, Hitchcock, and Wyckoff-Baird 1990). For additional discussions of the complexity of the tourism and land-use issues at /Xai/Xai and Gcwihaba, see Masilo-Rakgoasi (2002); Mbaiwa and Sakuze (2009) and Hitchcock et al. (2015).

Since the hunting ban was declared in Botswana in 2014, the Botswana government pulled back on its commitments to community-based natural resource management and the support of community trusts such as those of

Dobe, /Xai/Xai, and the Tsodilo Hills. As a result, funds available to community members from the lease fees of safari operators have dropped, and the number of jobs available has declined. The hunting ban had substantial negative effects on the well-being of San and other people in Botswana (Mbaiwa 2018; Coe 2020). Today, most of the benefits from tourism are going to private companies rather than to the community trusts themselves. The San and their neighbors are not happy with this privatization process, and they would like to see greater returns of the tourism benefits like they had in the past.

One of the concerns of tourists and local communities in southern Africa is the "shoot-to-kill" policies of the national defense forces, the police services, and personnel in the ministries of environment and tourism. This policy is aimed at poachers but also has impacts on local community members. Botswana has a stated "shoot-to-kill" policy which allows both police and the military to engage in what could be characterized as coercive conservation (for discussions of shoot-to-kill policies and their impacts, see Mongudhi, Konopo, and Ntibinyane 2016; Mogomotsi and Kefilwe 2017). Local people are deeply concerned about what they see as a trend toward militarization of conservation in Botswana as well as in Namibia and Zimbabwe (Duffy et al 2019; Hitchcock 2019b; Mogomotsi and Mogomotsi 2022). It is not only local people but also tourists who say that they have a certain amount of discomfort with seeing armed anti-poaching units in the places that they visit. Many tourists also say that they prefer to see local villagers going about their daily lives without fear of retribution for their normal activities from governments or non-government organizations. The preference of many tourists is to see "wild Africa" with both wild animals and local people living in their own ways without the restrictions of fences, border gates, anti-poaching patrols, and villagers running from them whenever they enter their communities.

## ZIMBABWE

Tourism in Zimbabwe is overseen by the Ministry of Environment, Climate, Tourism, and Hospitality Industry. The Zimbabwe Tourism Association (ZTA) also plays a role. Efforts have been made to expand tourism in Zimbabwe, especially in the northern and western portions of the country (Woyo and Woyo 2016, 2017, 2018). There are many private operators who take tourists to places of archaeological, cultural, environmental, and historical interest in Zimbabwe. None of these private safari companies or tourism operations are owned or managed by San, but they do sometimes employ San as guides.

Tshwa San, who number some 2,800 in western Zimbabwe (mainly in Tsholotsho District in Matabeleland North and *Bulilimamangwe* District in Matabeleland South), are seeking to expand tourist visits because of the potential they represent for providing income and employment (Ndlovu 2017, 2018; Ndlovu et al. 2022). The Tshwa are keen to take tourists to places in Hwange National Park and the Matopos Hills (Matobo National Park), where there are both archaeological sites dating to the Middle and Late Stone Age and Iron Age, as well as historically significant sites. Hwange National Park was originally established as a game reserve in 1928 and became a national park in 1961. Hwange has an area of 14,651 $km^2$ (5,657 $mi^2$) while Matobo National Park was established as a national park in 1926 (424 $km^2$ and 164 $mi^2$). Tshwa want to demonstrate some of their cultural practices, including traditional dances and craft manufacture, and serve as guides to the archaeological and historic sites.

Sometimes identifying themselves as "people of the two-way river," the Tshwa have utilized the resources of the savannas and riparian forested zones around the Nata (Manzanyama) River for hundreds of years in both Zimbabwe and Botswana. Unlike some other San groups in southern Africa, the Tshwa have totems. They also have part-time occupational specialists, including the following:

| | |
|---|---|
| káè-tcá-bá-tcò | hunter (sometimes termed dzimba) |
| tcóò-rà | healer (also nganga, which is a Kalanga loan) |
| ǁʔáé-xà | headman |

The Tshwa would like tourists to visit them so that they can discuss their cultural heritage and social, economic, and political systems with them (Ndlovu 2017, 2018; Hitchcock et al. 2018; Hitchcock, Begbie-Clench, and Murwira 2016; Hitchcock et al. 2018). They realize that tourists have much to offer in terms of providing sources of income, and they feel that the presence of tourists will instill cultural pride in their communities and encourage young people to learn more about Tshwa culture.

One type of tourism which interests tourists who come to Zimbabwe is so-called "dark tourism" which is tourism geared to visiting sites where disasters occurred or where massive human rights violations or genocide took place (see Lennon 2017; Kerr, Stone, and Price 2021). This is the case for those locations in Matabeleland where Tshwa, Kalanga, and Ndebele lost their lives in the period from 1981 to 1987 known as Gukurahundi ("sweeping away the chaff") which occurred while the new Zimbabwe government under President Robert Mugabe sought to rid the country of people he and his government defined as "dissidents" (Catholic Commission for Justice and Peace in Zimbabwe and Legal Resources Foundation 2008; Eppel 2014, 2015; Ngwenya 2018). Tshwa, among others in western Zimbabwe, wish to have memorials and commemorations in honor of the people who lost their lives during the time of "the troubles" (Ranger 1988, 1989).

Another type of tourism that Tshwa would like to engage in could be characterized as "agro-tourism" whereby Tshwa show tourists their gardens, agricultural fields, cattle and goat kraals, and residences. They feel that this kind of tourism would demonstrate to both tourists and Zimbabweans that they are not only hunter-gatherers but have a diversified lifestyle. Very proud of the strides they are making in terms of the education of the youth, Tshwa would also like tourists to see their schools, one example being the primary school at Mgodimasili.

Zimbabwe was, in many ways, the founder of community-based natural resource management in southern Africa through its establishment of the Communal Areas Management Program for Indigenous Resources (CAMPFIRE) (Martin 1986; Mberengwa 2000; Murembedzi 2001). However, some Tshwa have complained that areas set aside for CAMPFIRE-related purposes in Tsholotsho were limiting their collection of wild plants and firewood. Unfortunately, the benefit distribution to local communities has fallen off substantially over the years, with much of the money going to district councils (see, for example, Rihoy, Chirpzva, and Anstey 2010).

Tourism in Zimbabwe has also been affected by unfortunate events, such as the shooting of Cecil the lion by a Minnesota dentist engaged in a safari hunt in Tsholotsho in 2013 (see Lindsey et al. 2016), and the cyanide poisonings of elephants, rhinoceros, and other wild animals in Hwange National Park and Tsholotsho, also in 2013 (Muboko et al. 2016). Another factor that discourages tourist visits in Zimbabwe is the difficulty in obtaining currency, making it problematic for them to pay local people for products and activities. A major concern expressed by tourists in some of the protected areas of Zimbabwe was that they would hear gunshots and worry that this was because of poachers operating inside the parks (see Moyo and Phiri 2018 for a discussion of the impacts of the use of firearms in poaching in wildlife parks).

Archaeotourism and cultural tourism in Zimbabwe occur at Iron Age archaeological sites such as Domboshaba, Khami Ruins National Monument (a UNESCO World Heritage Site), Leopard's Kopje, and Great Zimbabwe. None of these sites employ San as guides at present. There were at least two San guides in the Natural History Museum in Bulawayo who were familiar with Tshwa San history and lifestyles. While the Tsoro-o-tso San Development Trust (TSDT) has sought to have the Zimbabwe San share benefits from tourism at archaeological and rock art sites, this has yet to be accepted by the Zimbabwe government. Funds generated by Zimbabwe national monuments currently go directly to the Zimbabwe government. Tshwa San have suggested having San cultural villages established at these archaeological sites where they could engage in tourism and sell San crafts along the lines of what is done in Namibia, but such a strategy has yet to be accepted by the Zimbabwe government.

One place where Tshwa San have sought to get visitation rights and benefit from protected area tourism is Hwange National Park, a protected area from which Tshwa were evicted in the latter part of the 1920s (Davison 1977, 1983; Haynes n.d.; Chingwe 2019). Relatively few tourists who visit Hwange National Park have the chance to see San, although ZimParks employs some San who work in Hwange (Gary Haynes, personal communication 2020). The cemetery at Main Camp in Hwange contains several graves of San guides who worked in the park for many years. Overall, Zimbabwe San benefit much less from tourism than the San in Botswana and Namibia, something that Zimbabwe San hope to change.

## CONCLUSIONS

Ecotourism represents an important part of the economies of Botswana, Namibia, and Zimbabwe. All three countries have produced detailed tourism plans. Most of these tourism plans are aimed at international (high-end) tourists because of the contributions that these kinds of tourists make to the foreign exchange earnings of the countries they visit.

Most San community members in the southern African region are in favor of tourism. At the same time, they want to have a greater say in how tourism is handled in their areas, for example, limiting the number of tourists who come there or dictating the necessity to clean up their camps/ They also want to see more CBT. While tourism has generated income for some local people and for their community-based organizations, the trickle-down effects of the income generated have often been relatively limited, and some of the benefits have been captured by community or private elites (for the example of Botswana CBT, see Rihoy and Maguranyanga 2010). Only South Africa offers the San a real chance to model their own tourism activities. They are the landowners of part of Kgalagadi Transfrontier Park, have land ownership in the Northern Cape, and not only own land in Platfontein with rock art which is visited by tourists but also have a joint project and program at !Khwa Ttu in the Western Cape.

CBT, while a very good idea, is generally not as beneficial to local communities as many of their members hope. A sizable proportion of the benefits of tourism at the community level goes to individuals who are educated and well connected with outside interests. The result is that equity, transparency, and accountability are key concerns of community members. Clearly, a challenge for sustainable tourism is coming up with a careful balance between optimization and maximization and spreading social and economic benefits among a broader range of people.

Table 3.3 presents data on tourist numbers visiting Botswana, Namibia, and Zimbabwe in 2017, 2018, and 2019. In general, the number of tourists visiting the three countries was on the increase. This trend was true until 2020,

**Table 3.3 Numbers of Tourists per Annum in Botswana, Namibia, and Zimbabwe**

| *Country* | *Tourists in 2017* | *Tourists in 2018* | *Tourists in 2019* | *Tourists in 2020* | *Tourists in 2021* |
|---|---|---|---|---|---|
| Botswana | 1,623,000 | 1,830,224 | 1,935,042 | 358,225 | 454,670 |
| Namibia | 1,608,018 | 1,659,762 | 1,762,079 | 196,026 | 354,508 |
| Zimbabwe | 2,450,000 | 2,579,927 | 2.200,000 | 639,000 | 375,799 |
| Total | 5,681,018 | 6,069,913 | 5,897,121 | 1,193,251 | 1,184,977 |

*Source*: Created by authors.
*Note*: Data obtained from the World Tourism Organization, Tourism Statistics Botswana, the Botswana Ministry of Environment, Natural Resources, Tourism, and Conservation, the Namibia Ministry of Environment, Forestry, and Tourism's Namibia Statistical Reports, the ZTA, the Zimbabwe National Statistical Agency, and the MECTH.

when the lockdowns due to the coronavirus pandemic led to a sharp decrease in the numbers of visitors.

Tourists say that two important factors that affect their tourism experience are the number of tourists in an area (they prefer that those numbers be small), and the cleanliness of villages and communities. If there is a lot of trash around, some tourists say, then the experience is not as good as it could be. A fairly large proportion of the tourism in southern Africa is not conducted in a sustainable manner according to tourists who visit the region. Tour companies and communities tend to push for having greater numbers of tourists, without much regard for the social, environmental, and economic impacts of these large numbers. There are relatively few examples of sustainable community tourism projects, although in some cases community-based organizations have attempted to set limits on the numbers of tourists visiting their areas and designating what they can do there. In the case of southern Africa, tourism has been used as a justification for removing Indigenous people from protected areas in all three of the countries with which we have dealt. (See Hitchcock et al. 2018; Sapignoli 2018; Chingwe 2019; Koot and Hitchcock 2019.)

Tourist benefits at the community level tend to be captured by men, leaving women, children, and the elderly in positions where they get relatively little. Indigenous people, by and large, benefit minimally from tourism except in rare cases. There are exceptions, however, as seen particularly in Namibia and, to a lesser extent, Botswana. In the case of Namibia, it is interesting to note that, as Naidoo et al (2016) point out, there are complementary benefits of tourism and hunting in communal conservancies such as Nyae Nyae and N=Jaqna. In other areas that either have low wildlife densities or do not allow hunting, the returns of tourism tend to be lower than those which have a mixture of hunting tourism and ecotourism. Such a strategy was not possible in Botswana after the hunting ban passed in 2014, which led to the closing down of numerous community trusts in the country (Mbaiwa 2018; Coe 2020), imposing huge costs on local communities.

Additional benefits, which San community members in Botswana, Namibia, and Zimbabwe see from tourism, include the ways that tourism can reinforce cultural identity and can help to underscore the importance and respect for their belief systems. This is evident in the appreciation of many tourists for the trance dances and other cultural activities of local community members. Many San want to engage in these activities for tourists not only to earn money, but also because they are a way to keep their values, customs, and traditions alive for themselves and their children.

## ACKNOWLEDGMENTS

An earlier version of this chapter was presented at the 78th annual meetings of the Society for Applied Anthropology, Philadelphia, Pennsylvania, USA, April 3–7, 2018. Funding for the research upon which this chapter is based was provided by the U.S. National Science Foundation (grant BCS 1122932), Brot für die Welt (Germany) (Project No. 2013 0148 G), and the Millennium Challenge-Account-Namibia (MCA-N). We wish to thank the Ministry of Environment, Forestry, and Tourism in Namibia, the Ministry of Environment, Natural Resources, Conservation, and Tourism in Botswana, and the Ministry of Environment, Water, and Climate and the Ministry of Local Government and Urban Development in Zimbabwe for permission to conduct this work. We also thank the Nyae Nyae Development Foundation of Namibia, the Nyae Nyae Conservancy, the N≠a Jaqna Conservancy, the /Xai/Xai Tlhabololo Trust, the Botswana National Museum office in the Tsodilo Hills, and the Tsoro-o-tso San Development Trust in Zimbabwe for all of their help. Megan Biesele, Flora Blommaert, Ben Begbie-Clench, Magdalena Broermann, Andy Chebanne, Tsamkxao Cique, Dave Cole, Roger Collinson, Lara Diez, Anne-Maria Fehn, Kelini Fernando, Judy Frost, Brian Jones, Ryan Klataske, Emanuel Koro, Steve/Kunta, Sam Makuvaza, Charity Manyeruke, Ashton Murwira, Kallie N!aice, Davy Ndlovu, Willem Odendaal, Aleksandra Ørbeck-Nilssen, Tsamkxao ≠Oma, Maria Sapignoli, Barbara Savage, Axel Thoma, Leon Tsamkxao, Catherine Willems, Nick Winer, Sarah Zungu, and the late Ignatius Mberengwa and the late Jack Hennessy provided us with useful insights on tourism and its impacts. The people of the Kalahari and the tourists who visit the Kalahari deserve special mention here for their willingness to share their ideas and experiences.

## REFERENCES

Andereck, K.L., K.M. Valentine, R.C. Knopf, and C.A. Vogt. "Residents' Perceptions of Community Tourism Impacts." *Annals of Tourism Research* 32 (2005): 10561076.

Angula, Hilma M., Greg Stuart Hill, David Ward, Greenwald Tatongo, Richard N. Diggle, and Robin Naidoo. "Local Perceptions of Trophy Hunting on Communal Lands in Namibia." *Biological Conservati*on 218 (2018): 26–31.

Begbie-Clench, Benjamin. *Illegal Grazing in Nyae Nyae Conservancy: Assessment and Recommendations*. Windhoek: Nyae Nyae Development Foundation of Namibia and Legal Assistance Centre and Tsumkwe: Nyae Nyae Conservancy, 2015.

Biesele, Megan and Robert K. Hitchcock. *The Ju/'hoansi San of Nyae Nyae and Namibian Independence: Development, Democracy, and Indigenous Voices in Southern Africa*. Paperback edition. New York: Berghahn Books, 2013.

Bollig, Michael, Robert K. Hitchcock, Cordelia Nduku, and Jan Reynders. *At the Crossroads: The Future of a Development Initiative. Evaluation of KDT (Kuru Development Trust), Ghanzi and Ngamiland Districts of Botswana*. Harare, Zimbabwe: Hivos Foundation, 2000.

Brondo, Keri Vacanti. *Land Grab: Green Neoliberalism, Gender, and Garifuna Resistance*. Tucson: University of Arizona Press, 2013.

Buurdeau, Laurent, Maria Gravari-Barbas, and Mike Robinson, eds. *World Heritage Sites and Tourism: Global and Local Relations*. London: Routledge, 2016.

Cassidy, Lin. *CBNRM and Legal Rights to Resources in Botswana*. CBNRM Support Program, Occasional Paper no. 4. Gaborone: IUCN SNV CRNRM Support Program, 2000.

Cassidy, Lin. "Power Dynamics and New Directions in the Recent Evolution of CBNRM in Botswana." *Conservation Science and Practice*, 3(1) (2021): e205. doi/org/10.1111/csp2206.

Catholic Commission for Justice and Peace in Zimbabwe and Legal Resources Foundation *Gukurahundi in Zimbabwe: A Report on the Disturbances in Matabeleland and the Midlands 1980 – 1988*. New York: Columbia University Press, 2008.

Chambers, Erve. *Tourism and Culture: An Applied Perspective*. Albany, NY: State University of New York Press, 1997.

Chingwe, Shuvai. "The San People of Zimbabwe: Decades After Land Dispossession." In *Displacement, Elimination, and Replacement of Indigenous People: Putting into Perspective Land Ownership and Ancestry in Decolonizing Contemporary Zimbabwe*, edited by Jairos Kangira, Artwell Nhemachena and Nelson Mlambo, 167–194. Bamenda, Cameroon: Langaa Research and Publishing CIG, 2019.

Coe, Katherine Kellam. An Exploration of the Adaptive Capacity of Community-Based Organizations in Northern Botswana in Response to A Hunting Ban. M.Sc. thesis, University of Montana, Missoula, Montana, 2020.

Cole, Dave *Water, Land and a Voice: The Nyae Development Foundation of Namibia. 36 Years of Support to the Ju/'hoan San Living in Namibia*. Windhoek: Nyae Development Foundation of Namibia, 2013.

Crick, Malcolm "Representations of International Tourism in the Social Sciences: Sun, Sex, Sights, Savings, and Servility." *Annual Review of Anthropology* 18 (1989): 307–344.

Davison, Ted. *Wankie: The Story of a Great Game Reserve*. Salisbury (Harare): Irwin Press, 1977.

Davison, Ted. *Wankie: The Story of a Great Game Reserve*. Third Edition. Bulawayo: Thorntree Press, 1983.

De Kadt, Emanuel. *Tourism: Passport to Development? Perspectives on the Social and Cultural Effects of Tourism in the Developing Countries*. Oxford: Oxford University Press, 1979.

Dube, Nqobizitha, Funa Moyo, Mkhokheli Sithole, Gracsious Ncube, Peter Nkala, Nevel Tshuma, Mandlenkosi Maphosa, and Clifford Mabhena. "Institutional exclusion and the tragedy of the commons: Artisanal mining in Matabeleland South Province, Zimbabwe. *The Extractive Industries and Society* 3(4) (2016): 1084–1094.

Duffy, Rosaleen, Francis Massé, Emile Smidt, Esther Marijnen, Bram Büscher, Judith Verweijen, Maano Ramutsindela, Trishant Simlai, Laure Joanny, and Elizabeth Lunstrum. "Why We Must Question the Militarization of Conservation." *Biological Conservation* 232 (2019): 66–73.

Eppel, Shari. "'Bones in the Forest' in Matabeleland, Zimbabwe: Exhumation as a Tool for Transformation." *International Journal of Transitional Justice* 8(3) (2014): 404–425.

Eppel, Shari. "The Heroic and the Hidden Dead: Zimbabwe and Exhumations. In *Forensic Archaeology: A Global Perspective*, edited by W.J. Mike Groen, Nicholas Marquez-Grant, and Robert C. Janoway, 359–368. New York: John Wiley and Sons, 2015.

Fennell, David. *Ecotourism*. Fifth Edition. London and New York: Routledge. 2020

Giraudo, Rachel F. "World Heritage, Tourism Development, and Identity Politics at the Tsodilo Hills." In *World Heritage Sites and Tourism: Global and Local Relations*, edited by Laurent Buurdeau, Maria Gravari-Barbas, and Mike Robinson, 77–93. London: Routledge, 2015.

Giraudo, Rachel and Benjamin W. Porter. "Archaeotourism and the Crux of Development." *Anthropology News* (2010): 7–8.

Government of Botswana and United Nations Development Program. *Managing the human-wildlife interface to sustain the flow of agro-ecosystem services and prevent illegal wildlife trafficking in the Kgalagadi and Ghanzi Drylands, Botswana*. Gaborone: Government of Botswana and United Nations Development Program, 2021.

Green, Isobel and Jarkko Saarinen. "Changing Environment and the Political Ecology of Authenticity in Heritage Tourism: A Case of the Ovahimba and the Ju/'Hoansi-San Living Museums in Namibia." In *Southern African Perspectives on Sustainable Tourism Management. Geographies of Tourism and Global Change*, edited by Saarinen, J., Lubbe, B., Moswete, N.N. Cham: Springer, 2020.

Guenther, Mathias. "Ethno-Tourism and the Bushmen." In *Self- and Other-Images of Hunter-Gatherers*, edited by Henry Stewart, Alan Barnard, and Deeichi Omura, 47–64. Senri Ethnological Studies No. 60. Osaka, Japan: National Museum of Ethnology, 2002.

Haratsebe, Maria, Naomo Moswete, and Jarkko Saarinen, eds. *Cultural Tourism in Southern Africa*. Bristol, UK: Channel View Publications, 2016.

Hays, Jennifer. "The Invasion of Nyae Nyae: A Case Study in on-going Aggression against Hunter-gatherers in Namibia." In *Forum Conference on Indigenous*

*Peoples 2009: Violent Conflicts, Ceasefires, and Peace Accords Through the Lens of Indigenous Peoples,* 25–32. Tromso: Forum for Development Cooperation with Indigenous Peoples, 2009.

Hays, Jennifer, and Robert K. Hitchcock. "Land and Resource Rights in the Tsumkwe Conservancies – Nyae Nyae and N≠a Jaqna." In *"Neither Here nor There": Indigeneity, Marginalisation, and Land Rights in Post-Independence Namibia*, edited by Willem Odendaal and Wolfgang Werner, 183–214. Windhoek: Legal Assistance Center, 2020.

Haynes, Gary. *Hwange National Park: The Forest with a Desert Heart*. Manuscript in author's possession, University of Nevada-Reno, Reno, Nevada, USA. 2013.

Hitchcock, Robert K. "Cultural, Economic, and Environmental Impacts of Tourism among Kalahari Bushmen." In *Tourism and Culture: An Applied Perspective*, edited by Erve Chambers, 93–128. Albany, NY: State University of New York Press, 1997.

Hitchcock, Robert K. "Refugees, Resettlement, and Land and Resource Conflicts: The Politics of Identity among !Xun and Khwe San of Northeastern Namibia." *African Study Monographs* 33(2) (2012): 73–132.

Hitchcock, Robert K. *Improving the Viability and Sustainability of the Nyae Nyae Conservancy and the N≠a Jaqna Conservancy: A Mid-Term Evaluation of an NNDFN/Brot für die Welt Project in Namibia.* Windhoek: Nyae Nyae Development Foundation of Namibia and Berlin: Brot für die Welt, 2015.

Hitchcock, Robert K. "Ecotourism, Heritage Tourism and Community-based Development in a Southern African Borderland." In *Bewildering Borders: The Economy of Conservation,* edited by Werner Zips and Manuela Mairitsch-Zips, 61–89. Münster: Lit Verlag and Wien: Lit Verlag Wien, 2019a.

Hitchcock, Robert K. "The Impacts of Conservation and Militarization on Indigenous Peoples: A Southern African San Perspective." *Human Nature* 30 (2019b): 217–241

Hitchcock, Robert K. "The Plight of the Kalahari San: Hunter-gatherers in a Globalized World." *Journal of Anthropological Research* 76(2)(2020): 164–-184.

Hitchcock, Robert K., Ben Begbie-Clench, Davy Ndlovu, Ashton Murwira, and Ignatius Mberengwa. "Land, Livelihoods, and Empowerment Among the San of Western Zimbabwe. In *Research and Activism Among the Kalahari San Today: Ideals, Challenges, and Debates*, edited by R. Fleming Puckett and Kazunobu Ikeya, 251–282. Senri Ethnological Studies 99. Osaka: National Museum of Ethnology, 2018.

Hitchcock, Robert K., Benjamin Begbie-Clench, and Ashton Murwira. *The San in Zimbabwe: Livelihoods, Land, and Human Rights*. Report No. 22. Copenhagen: International Work Group for Indigenous Affairs and Johannesburg: Open Society Initiative for Southern Africa, 2016.

Hitchcock, Robert K. and Rodney L. Brandenburgh. "Tourism, Conservation, and Culture in the Kalahari Desert, Botswana." *Cultural Survival Quarterly* 14(1) (1990): 1–5.

Hitchcock, Robert K. and Judith Frost. (2021). Botswana. In *Indigenous World 2021*, edited by Dwayne Mamo, 37–47. Copenhagen: International Work Group for Indigenous Affairs.

Hitchcock, R.K., M. Sapignoli, M, Main, and W.A. Babchuk. "The Politics and Economics of Community-Based Natural Resource Management in /Xai/Xai, Ngamiland, Botswana." *African Study Monographs* 36(4) (2015): 211–260.

Hulme, David and Marshall Murphree. "Communities, Wildlife and the 'New Conservation' in Africa." *Journal of International Development* 11 (1999): 277–285.

Hüncke, Anna and Stasja Koot. "The Presentation of Bushmen in Cultural Tourism: Tourists' Images of Bushmen and the Tourism Provider's Presentation of (Hai//om) Bushmen at Treesleeper Camp, Namibia." *Critical Arts: South-North Cultural and Media Studies* 26(5) (1999): 671–689.

Hunter, Malcolm L., Robert K. Hitchcock, and Barbara Wyckoff-Baird. "Women and Wildlife in Southern Africa." *Conservation Biology* 4(4) (1990): 448–451.

Jones, Brian. *Community-based Natural Resource Management in Botswana and Namibia: An Inventory and Preliminary Analysis of Progress*. Windhoek: Southern African Sustainable Use Specialist Group and London: International Institute for Environment and Development, 1997.

Jones, Brian and Marshall Murphree. "The Evolution of Policy on Community Conservation in Namibia and Zimbabwe." In *African Wildlife and Livelihoods: The Promise and Performance of Community Conservation*, edited by David Hulme and Marshall Murphree, 38–57. London: James Currey, 1997.

Keaketswe, Kabo " Staircase Makes Gcwihaba Accessible." *Daily News*, 27 November 2017.

Keitumetse, Susan and Olivia Nthoi. "Investigating the Impact of World Heritage Tourism on the Intangible Heritage of a Community, Tsodilo Hills World Heritage Site, Botswana." *International Journal of Intangible Heritage* 4 (2009): 144–150.

Kerr, Mary Margaret, Philip Stone, and Rebecca H. Price. "Young Tourists' Experiences at Dark Tourism Sites: Towards a Conceptual Framework." *Tourist Studies* 21(2) (2021): 198–218.

Koot, Stasja P. *Dwelling in Tourism: Power and Myth amongst Bushmen in Southern Africa*. African Studies Collection Volume 54. Leiden, The Netherlands: African Studies Centre, 2013.

Koot, Stasja and Robert K. Hitchcock. "In the Way: Perpetuating Land Dispossession of the Indigenous Hai//om and the Collective Action Lawsuit for Etosha National Park and Mangetti West, Namibia." *Nomadic Peoples* 23(1) (2019): 55–77.

Koro, Emmanuel. COVID-19 Plunges Zimbabwe into Wildlife Management Crisis. *Chronicle*, 9 July 2020.

Lee, Richard B. *The !Kung San: Men, Women, and Work in a Foraging Society*. Cambridge: Cambridge University Press, 1979.

Lee, Richard B. *The Dobe Ju/'hoansi*. Fourth Edition. Independence, Kentucky: Cencage, 2013.

Lennon, John. "Dark Tourism." In *Oxford Research Encyclopedia of Criminology and Criminal Justice*. 2017. doi:10.1093/acrefore/9780190264079.013.212

Lindsey, P.A., P.A. Roulet, and S.S. Romañach. "Economic and Conservation Significance of the Trophy Hunting Industry in Sub-Saharan Africa." *Biological Conservation* 134 (2007): 455–469.

Lindsey, Peter A., Guy A. Balme, Paul J. Funston, Philipp H. Henschel, and Luke T.B. Hunter. "Life after Cecil: Channeling Global Outrage into Funding for Conservation in Africa." *Conservation Letters* 9(4) (2016): 296–301.

Marshall, John. *A Kalahari Family* (video). Watertown: Documentary Educational Resources, 2003.

Marshall, Lorna. *The !Kung of Nyae Nyae*. Cambridge: Harvard University Press, 1976.

Martin, Rowan B. *Communal Areas Management Programme for Indigenous Resources (CAMPFIRE)*. Harare, Zimbabwe: Branch of Terrestrial Ecology, Department of National Parks and Wildlife Management (DNPWLM), 1986.

Masilo-Rakgoasi, Rosinah. An Assessment of the Community-Based Natural Resource Management Approach and Its Impact on the Basarwa: Case Study of / Xai/Xai and Gudigwa Communities. M.A. thesis in Development Studies, University of Botswana, Gaborone, Botswana, 2002.

Mbaiwa, Joseph E. Tourism Development, Rural Livelihoods, and Conservation in the Okavango Delta, Botswana. PhD dissertation, Texas A&M University, College Station, Texas, USA, 2008.

Mbaiwa, Joseph E. "Hosts and Guests: Stereotypes and Myths of International Tourism in the Okavango Delta." In *African Hosts and their Guests: Cultural Dynamics of Tourism,* edited by Walter van Beek and Annette Schmidt, 117–136. London: James Currey, 2012.

Mbaiwa, Joseph. "The Commodification of World Heritage Sites: The Case Study of Tsodilo Hills in Botswana." In *Cultural Tourism in Southern Africa*, edited by Maria Haratsebe, Naomo Moswete, and Jarkko Saarinen, 101–120. Bristol, UK: Channel View Publications, 2016a.

Mbaiwa, Joseph. "Tourism Development, Dispossession, and Displacement of the Communities in the Okavango Delta, Botswana." In *Political Ecology of Tourism*, edited by Sanjay Nepal and Jarkko Saarin, 193–206. London: Routledge, 2016b.

Mbaiwa, Joseph E. Poverty or Riches: Who Benefits from the Booming Tourism Industry in Botswana? *Journal of Contemporary African Studies* 35(1)(2017): 93–112.

Mbaiwa, Joseph E. Effects of the Safari Hunting Tourism Ban on Rural Livelihoods and Wildlife Conservation in Northern Botswana." *South African Geographical Journal* 100(1)(2018): 41–61.

Mbaiwa, Joseph E. and Wame L. Hambira. "Enclaves and Shadow State Tourism in the Okavango Delta, Botswana." *South African Geographical Journal* 102(1) (2020): 1–21.

Mbaiwa, J. and L. Sakuze. "Cultural tourism and livelihood diversification: The case of Gcwihaba Caves and XaiXai Village in the Okavango Delta, Botswana. *Journal of Tourism and Cultural Change* 7(1) (2009):61–75.

Mberengwa, Ignatius. The Communal Areas Management Program for Indigenous Resources (CAMPFIRE) and Rural Development in Zimbabwe's Marginal Areas:

A Study in Sustainability. Ph.D. Dissertation, University of Nebraska, Lincoln, Nebraska, 2000.

Mogomotsi, E.J. and Patricia Kefilwe. "Live by the Gun, Die by the Gun: Botswana's 'Shoot to Kill' Policy as an Anti-poaching Strategy." *South Africa Crime Quarterly* 60 (2017): 51–59.

Mogomotsi, Goemeone E. J. and Patricia K. Mogomotsi. "Law of Armed Conflict in Non-International Hostilities: The Militarisation of Wildlife Conservation in Africa." *Journal of International Wildlife Law & Policy*, 2022. doi: 10.1080/13880292.2021.2059239

Mongudhi, Tileni, Joel Konopo, and Ntibinyane Ntibinyane. "Deadly borders ... 30 Namibians killed through Botswana's shoot-to-kill policy." *The Namibian*, 9 March 2016.

Moyo, Wisdom and Yvonne Phiri. "The Impact of the Use of Firearms in Poaching in Wildlife Parks." *Saudi Journal of Humanities and Social Sciences* 3(8) (2018): 973–978.

Muboko, Never, Edson Gandiwa, Victor Muposhi, and Tawanda Tarakini. Illegal Hunting and Protected Areas: Tourist Perceptions on Wild Animal Poisoning in Hwange National Park, Zimbabwe." *Tourism Management* 52 (2016): 170–172.

Murembedzi, James. "Committees, Rights, Costs, and Benefits: Natural Resource Stewardship and Community Benefits in Zimbabwe's CAMPFIRE Program." In *African Wildlife and Livelihoods: The Promise and Performance of Community Conservation*, edited by David Hulme and Marshall Murphree, 244–255. London: James Currey, 2001.

Naidoo, Robin, L. Chris Weaver, Richard W. Diggle, Greenwell Matongo, Greg Stuart-Hill, and Chris Thouless. "Complementary Benefits of Tourism and Hunting to Communal Conservancies in Namibia." *Conservation Biology* 30(3) (2016): 628–638.

Ndlovu, Davy. *A New Age for the San in Zimbabwe*. Harare: Bhabhu Books, 2017.

Ndlovu Davy. *My Culture, My Pride: Reclaiming the Tjwa Cultural Identity*. Tsholotsho: Tsoro-o-tso San Development Trust, 2018.

Ndlovu, Davy. *A Case Study on How COVID-19 Lockdown is Impacting Continued Access to Education for Indigenous Minority San Students in Tsholotsho*. Bulawayo: Tsoro-o-tso San Development Trust, 2020.

Ndlovu, Davy, Ben Begbie-Clench and Robert K. Hitchcock. "Zimbabwe." In *The Indigenous World 2021*, edited by Dwayne Mamo, 165–172. Copenhagen: International Work Group for Indigenous Affairs, 2021.

Ndlovu, Davy, Ben Begbie-Clench, Robert K. Hitchcock, and Melinda C. Kelly. "The Tshwa San of Zimbabwe: Land, Livelihoods, and Ethnicity." In *Livelihoods of Ethnic Minorities in Rural Zimbabwe*, edited by Kirk Helliker, Patience Chadambuka, and Joshua Matanzima, 31–50. Cham, Switzerland: Springer. 2022.

Ngwenya, Dumisani. ed. *Healing the Wounds of Gukurahundi in Zimbabwe: A Participatory Action Project*. Dordrecht: Springer. 2018.

Oma, Kxao Moses and Axel Thoma. Does Tourism Support or Destroy the Indigenous Cultures of the San? Paper presented at the Workshop on Tourism and Indigenous Peoples, Geneva, Switzerland, July 28, 1998.

Owens, Mark and Delia Owens. *Preliminary Final Report on the Central Kalahari Predator Research Project*. Report to the Department of Wildlife and National Parks, Gaborone, Botswana, 2001

Owens, Mark and Delia Owens. *Cry of the Kalahari*. Boston, MA: Houghton-Mifflin, 1984.

Ranger, Terence O. "War, Violence, and Healing in Zimbabwe." *Journal of Southern African Studies* 18(3) (1988): 698–707.

Ranger, Terence O. "Matabeleland since the Amnesty." *African Affairs* 88 (351) (1989): 161–173.

Republic of Botswana. *Tourism Policy*. Government Paper No. 2 of 1990. Gaborone, BW: Government Printer, 1990.

Republic of Botswana. *National Ecotourism Policy*. Gaborone, BW: Government Printer. 1991

Republic of Botswana. *Wildlife Conservation and National Parks Act, 1992*. Act No. 28 of 1992. Gaborone, Botswana, Government Printer, 1992.

Republic of Botswana. *Tourism Act 2009*. Gaborone, BW: Republic of Botswana, 2009a.

Republic of Botswana. *Tourism Organisation Act 2009*. Gaborone, BW: Republic of Botswana, 2009b.

Republic of Namibia. *Namibia Tourism Board Act, 2000*. Windhoek; Government of the Republic of Namibia, 2000.

Republic of Namibia. *Environmental Management Act*. Act No. 7 of 2007. Windhoek: Ministry of Environment and Tourism and Government of Namibia, 2007

Republic of Namibia. *National Policy on Tourism for Namibia*. Windhoek: Ministry of Environment and Tourism and Government of the Republic of Namibia, 2008

Republic of Zimbabwe. *National Tourism Policy*. Harare: Republic of Zimbabwe, 2016.

Rihoy, Liz and Brian Maguranyanga. "The Politics of Community-Based Resource Management in Botswana." In *Community Rights, Conservation, and Contested Lands: The Politics of Natural Resource Governance in Africa*, edited by Fred Nelson, 55–78. London and Washington D.C.: Earthscan, 2010.

Rihoy, Liz, Chaka Chirpzva, and Simon Anstey. 'People are Not Happy': Crisis, Adaption, and Resilience in Zimbabwe's CAMPFIRE Program." In *Community Rights, Conservation, and Contested Land: The Politics of Natural Resource Governance in Africa*, edited by Fred Nelson, 174–201. London and Washington, D.C.: Earthscan, 2010.

Ritterband, Salomé. *Tracking Indigenous Heritage: Ju/'hoansi San Learning, Interpreting, and Staging Tradition for a Sustainable Future in Cultural Tourism in the Tsumkwe District of Namibia*. Zurich and Wien: Lit Verlag GmbH and Company, 2018.

Rozemeijer, Nico, ed. *Community-Based Tourism in Botswana: The SNV Experience in Three Community Tourism Projects*. Gaborone, BW: SNV (Netherlands Development Organization), 2001.

Sapignoli, Maria. *Hunting Justice: Displacement, Law, and Activism in the Kalahari*. Cambridge: Cambridge University Press, 2018.

Schuster, B. *Proceedings of the 4th National CBNRM Conference in Botswana and the CBNRM Status Report, 20–23 November 2006*, IUCN Botswana, Gaborone, Botswana, 2007.

Snyman, Sue and Kelly S. Bricker. "Living on the Edge: Benefit Sharing from Protected Area Tourism." *Journal of Sustainable Tourism* 27(6) (2019): 705–719.

Stadler, Simon, Director. *Ghostland: The View of the Ju/'hoansi*. Video. New York and Montreal: *Cargo Film & Releasing,* 2016.

Stonich, Susan. "Political Ecology of Tourism." *Annals of Tourism Research* 25(1) (1998): 25–54.

Streicher, Daniel G. and Amy T. Gilbert. "Contextualizing Bats as Viral Reservoirs. *Science* 370 (2020): 172–173.

Sufu, Eyassu, Ayana Anyassa and W.S. Boitumilo. "Community-Based Camel Ecotourism in Botswana: Current Status and Future Perspectives. *Journal of Camelid Science* 48 (2018): 118–133.

Taylor, Michael. "The Politics of Cohabitation: Social History of Tsodilo." In *The Tsodilo Hills: Copper Bracelet of the Kalahari*, edited by Alec Campbell, Larry Robbins, and Michael Taylor, 116–125. East Lansing: Michigan State University Press and Gaborone: Botswana Society, 2010.

Taylor, Michael. 'We are not taken as people': Ignoring the Indigenous Identities and History of Tsodilo Hills World Heritage Site, Botswana. In *World Heritage Sites and Indigenous Peoples' Rights*, edited by Stefan Disko and Helen Tugendhat, 119–129. Copenhagen: International Work Group for Indigenous Affairs, 2014.

Thomsen, Jennifer Marie, Selma Lendelvo, Katherine Coe and Melanie Rispel. "Community Perspectives of Empowerment from Trophy Hunting Tourism in Namibia's Bwabwata National Park." *Journal of Sustainable Tourism* (2021). doi: 10.1080/09669582.2021.1874394

Tomaselli, Keyan G. Another Look at John Marshall and the Ju/'hoansi: Reassessing a Documentary Lifetime. *Visual Anthropology* 34(2) (2021): 101–133.

United Nations Development Program and Government of Botswana *Social Safeguards in the Ghanzi-Kgalagadi Drylands Ecosystem Area*. Gaborone: United Nations Development Program and Government of Botswana. 2021.

United Nations World Tourism Organization. *Tourism: The year 2000 and beyond qualitative aspects*. Madrid: United Nations World Tourism Organization, 1993.

United Nations World Tourism Organization. *Global Report on Women in Tourism*. Second Edition. Madrid: United Nations World Tourism Organization, 2019.

Van Der Post, Laurens. *The Lost World of the Kalahari*. (film) London: British Broadcasting Corporation, 1958.

Van Der Wulp, Christa. Transformation of Communal Lands: Illegal Fences in the N≠a Jaqna Conservancy. M.A. Thesis, Wageningen University, Wageningen, The Netherlands, 2016.

Van der Wulp, Christa and Paul Hebinck. "Fighting Fences and Land Grabbers in the Struggle for the Commons in N≠a Jaqna, Namibia." *African Affairs* 120 (2021): 417–443.

Veerman, Jan, Abhishek Kumar, and Deepak R Mishra. "Exceptional Landscape-wide Cyanobacteria Bloom in Okavango Delta, Botswana in 2020 Coincided with a Mass Elephant Die-off E." *Harmful Algae* 2021. 111. doi:10.1016/j.hal.2021.102145.

Welch, Cameron. '*Land is Life, Conservancy is Life": The San and the N≠a Jaqna Conservancy, Tsumkwe District West, Namibia*. Basel Namibia Studies Series. Basel, Switzerland: Basler Afrika Bibliographen, 2018.

World Bank. *Supporting Sustainable Livelihoods through Wildlife Tourism*. Washington, D.C.: World Bank Group, 2018.

World Bank. *Tools and Resources for Nature Based Tourism*. Washington, D.C.: The World Bank, 2020a.

World Bank. *Banking on Protected Areas: Promoting Sustainable Protected Area Tourism to Benefit Local Communities*. Washington, D.C.: The World Bank, 202b.

Woyo, Erisher and Edith Woyo. "The Impacts of Tourism Development at Domboshava National Monument." *International Journal of Research in Tourism and Hospitality* 2(3) (2016): 29–36.

Woyo, Erisher and Edith Woyo. "An Assessment of the Factors Influencing Visitation to Cultural Sites in Northern Zimbabwe." *Journal on Tourism & Sustainability* 1(1) (2017): 71–91.

Woyo, Erisher and Edith Woyo. "Towards the Development of Cultural Tourism as an Alternative for Tourism Growth in Northern Zimbabwe." *Journal of Cultural Heritage Management and Sustainable Development* (2018). doi:10.1108/JCHMSD-08-2016-0048

World Travel and Tourism Council. *Botswana: The impacts of travel and tourism on jobs and the economy*. Madrid: World Travel and Tourism Council, 2017.

Zhou, Zibanai. "Human–wildlife tourism conflict: Towards an inclusive wildlife resources utilisation approach in southern Africa. *GeoJournal* (2022). doi:10.1007/s10708-022-10667-8

Zips, Werner. he Ambiguous Art of Entertainment: Living Museums and Cultural Villages in Namibia. Paper presented in a symposium titled "Thriving Future: Community-Based Research and Planning", Conference on Hunter-Gatherer Studies (CHAGS) 11, Vienna, Austria, 7–11 September 2015.

Zips, Werner and Manuela Mairitsch-Zips. *The Ambiguous Art of Ethnoentertainment: Living Museums and Cultural Villages in Namibia*. Video, Vienna (Wien): Universitat Wien, Austria, 2015.

Zips, Werner and Manuela-Mairitsch-Zips, eds. *Bewildering Borders: The Economy of Conservation*. Münster: Lit Verlag and Wien: Lit Verlag Wien, 2019.

*Chapter 4*

# Ecotourism, Anthropology, and San Community Docents and Activists

## *Some Interactions*

Megan Biesele

In the apartheid and early post-apartheid periods in southern Africa, many community development, advocacy, and educational projects involving San peoples were rightly criticized for top-down policies and practices. More recently, pushback from San groups and a liberalizing international development climate have combined to move southern African activist efforts with and on behalf of San groups in more productive, consultative, and collaborative directions. Contradictions, as always, abound, but substantial positive changes in community and organizational relationships have occurred, often informed by applied anthropological research and planning. Satisfactory outcomes and social atmospheres more conducive to realistic problem-solving have begun to evolve in some advocacy and activist contexts. I believe this is largely because donors and outside implementers of projects have begun to listen better to local concerns, react more favorably to local initiatives, and respect the resilience and wisdom of Indigenous individuals and groups.

Much of this wisdom revolves around the practice of egalitarian values, widely described in anthropological literature on the San (e.g., Biesele and Hitchcock 2013, Lee and Daly 1999, Lee and DeVore 1976). These values are often, amazingly, not generally held by outsiders to San society, even those who profess to want to help. Realizing in recent decades that outsiders had and would continue to threaten their rights and their way of life through sheer ignorance of community values, the San began to use media (film, local and national radio, and anthropological publications and newspapers) to publicize their community projects, intentions, and deeply-held social practices and beliefs. The social values of egalitarianism include pervasive sharing according to mutually understood rules; great attention to matters of equity and inclusiveness; and equal respect given to all, that is, no sexism, ageism, or discrimination on the basis of ability or personality. Part of this chapter is devoted to a recent exercise in promoting knowledge of such San values initiated by Naro San people living near D'kar, Botswana (e.g., Kuru Family of Organizations, 2012, 2014). I would also like to draw attention to a report that came out in the journal *Science* on March 17, 2017, about a comprehensive new San code of ethics for researchers (Nordling 2017). This effort was undertaken by San organizations partly because "approval by university research ethics committees is not sufficient to comply with (our) code," according to Leana Snyders of the South African San Council in Upington, South Africa, which helped to create the code (Nordling 2017, Gosling 2017, South African San Institute 2017).

This enlightenment has not, of course, yet penetrated the whole spectrum of organizations claiming to exist for the betterment of San and other Indigenous peoples. Too often, applied anthropologists and community activists must spend a disproportionate part of their time and energy doing "damage control" due to ill-conceived initiatives brought by outside entities with intentions ranging from benign but unwise to frankly exploitative and paternalistic projects. Often the productive, community-based projects are not publicized precisely because individuals and researchers involved in them are too busy carrying them out with integrity while avoiding competing, deleterious projects. Then, too, there is the issue of publication capacity and cost, which must of necessity be of lower priority to local workers and is often left up to researchers who may or may not find appropriate outlets for the information.

In partial remedy, this paper intends to shed light on several recent, illuminating examples of favorable interactions among (1) San community ecotourism docents and activists in Namibia, Botswana, and South Africa; (2) the tourists and donors they task themselves to educate and influence; and (3) applied anthropologists and their publications on the communities in question. The word "docent" is used in this context more generally than the way it is used in the United States, where it means a tour leader at a museum. For purposes of this chapter, "docents" may be everything from ecotour guides to crafts instructors to expert local animal behaviorists, trackers, and even healers who interact with a tourist public.

My examples include, first, an ecotourism course given jointly by the Working Group of Indigenous Minorities in Southern Africa (WIMSA) and The University Centre for Studies in Namibia (TUCSIN) for San community docents held in Windhoek, Namibia in 2002–2003 (TUCSIN 2004, 16). Second, it covers sets of values guidelines created by San activists, starting in 2008 and ongoing today with four, eventually to include eight, language groups, to inform San community development and ecotourism projects in Botswana and Namibia (e.g., Kuru Family of Organizations, 2012, 2014). Third, it describes recent (2012–2020) updates to docent education for ecotours at !Khwa ttu San Culture and Education Centre, Yzerfontein, South Africa (!Khwa ttu: San Spirit Shared, n.d). The examples share both increasing use and increasing critique by San docents and activists of written resources on the relevant communities as furnished by anthropologists. They suggest ways San community members may continue to increase their control of community representation decisions. Importantly, they highlight successful approaches to Indigenous ecotourism initiatives that may be of use in other parts of the world.

## LET'S COMMUNICATE! SAN PUBLIC RELATIONS OFFICERS COURSE 2002–2003, WINDHOEK, NAMIBIA

TUCSIN and WIMSA collaborated in 2002–2003 to offer a course in public relations for twelve San students (six men, six women) from all over Namibia and parts of Botswana (TUCSIN 2004, 16). Most of the students were docents or were training to be docents or other sorts of spokespersons for their communities. This course was a new departure for both then twenty-five-year-old pioneering educational institution TUCSIN and then six-year-old activist working group WIMSA.

An informal needs assessment conducted among the WIMSA member organizations and the San Traditional Authorities had established that more San should be enabled to acquire the organizational, administrative, linguistic, and promotional skills needed to represent their communities at conferences, Indigenous peoples' fora, seminars, workshops, and so on. At the launch of the Regional Assessment of the Status of the San in Southern Africa by the Namibian Legal Assistance Center in late 2001, TUCSIN and WIMSA decided to implement a project aimed at training young San with some working experience in community development to represent their interests at local, regional, national, and international levels. Dr. Beatrice Sandelowsky, an archaeologist and TUCSIN's founder, designed a year-long course consisting of four, six to eight-week sessions of classwork interspersed with long (four to six weeks each) breaks for the practical application of skills acquired during the periods of study on the TUCSIN campus in Windhoek.

Tutorials in English, Southern African History, Cultural Anthropology, Accounting, and Law were offered by a large group of international lecturers speaking on a volunteer basis. Many of these lecturers, myself included, were visiting Namibia in connection with their own projects and research. The course was designed to benefit from their presence in the country and to bring together the community activist, educational, and research communities working with Indigenous San communities. Local educators volunteered their help to teach a wide range of skills, including negotiation, leadership, accounting, public speaking, etiquette, and public relations.

The course started with a four-day excursion to the well-known rock art sites of Brandberg and Twyfelfontein in Namibia. (Many of the students had never had a chance to visit these national monuments, whose painted images are attributed by archaeologists to their San ancestors.) During the first term, WIMSA Coordinator Axel Thoma and his San counterpart Joram /Useb gave lectures on San intellectual property rights, donor requirements in project funding, and WIMSA's Oral Testimony Collection Project. One of the practical exercises, which the group as a whole was tasked to perform together, consisted of staging a press conference on community concerns. The press conference was screened on national TV in Namibia and written up in the daily newspapers in several national languages.

The second term highlight was a course in "Speechcraft." It covered effective listening, body language, and the preparation, presentation, and evaluation of speeches. Issues for performative role-playing exercises in the third term were chosen by the students themselves and included HIV/AIDS, "Taking law into our own hands," and "Leadership and corruption." The fourth term focused on computer skills and evaluations by the students of their own progress, the lecturers' performance, and the overall curriculum.

I participated, along with educator Melissa Heckler, in one of the sessions of this Public Relations Course; in it, students critiqued representations of San in anthropological literature. This was a real eye-opener for me, as the class got down very quickly to some (pretty uncomfortable) brass-tacks questions. "Why is the healer shown lying down in a loincloth at a healing dance in the book you wrote with Richard Katz and Verna St. Denis, p. 19?" (Katz, Biesele, & St. Denis, 1997). "Wouldn't it have been better public relations to choose a different picture, one with him upright and not wearing skins?" "How was the decision made which pictures to put into the book?" There were lively debates over all these questions, with the giddy sense prevailing that these were questions people had been itching to bring up but had never until now found the right forum. We all learned a lot that day.

Though the important organization WIMSA is now, sadly, defunct, TUCSIN hopes to give the course again when possible; there have been many requests from other groups involved in ecotourism in Namibia, especially the Himba. Since that time, TUCSIN, active throughout Namibia for the last thirty-five years (TUCSIN 2013), has had over 15,000 Namibian students in its courses. It is responsible, at last count, for the fact that fully 1,400 Namibians have qualifications beyond the B. A., a figure undreamt of during the early apartheid years of the country. In recent years, TUCSIN has undertaken a new project focusing solely on San people. Called TUCSIN-Tsumkwe, it is a multipurpose educational and training project in the Ju|'hoan area of Nyae Nyae, northern Namibia. After the last four years of extensive data gathering and community input, TUCSIN has now approved a strategic plan aimed at addressing the issues that emerged. TUCSIN-Tsumkwe has committed to implementing programs that address issues of income generation, academic success, and cultural preservation.

Income generation has initially focused on access to revenue from traditional skills. Tourism is being expanded in a respectful and authentic way by hiring "elder teachers" to share their cultural knowledge and skills within the community. Some vocational training opportunities have been created, but employment opportunities are TUCSIN's main concern now. Academic programs include support to children in school by supplying basic needs (toiletries and nutrition), after-school programs and activities, and helping parents prepare their children for the formal education system. Cultural programs thus far link the elder teachers with school children to share their knowledge and stories. TUCSIN is also talking with a new partner, Khwa ttu in South Africa (discussed below), about becoming a satellite center for their museum. This project is intended to empower communities to document and archive the past and present and to make the records accessible to the whole community.

## SAN VALUES: 2008 AND ONGOING. KURU FAMILY OF ORGANIZATIONS, D'KAR, BOTSWANA

Members of the Ncoakhoe (San) groups of Botswana had long realized that outside forces with which they were increasingly contending to hold onto their land and resource rights did not know or respect the values by which they lived. Accordingly, Naro leaders James Tshabu Morris and Aaron Johannes began in 2008 to not only make their San-owned, local Kuru Family of Organizations (KFO) a values-driven NGO but to carry the process forward to other groups as well. The first San values documentation process took place, with facilitation by Braam and Willemien LeRoux, Kuru's founders, in 2008–2009. A Namibian NGO, Integrated Rural Development and Nature Conservation, helped support the documentation and integration of San values for leadership development. KFO regarded San values as tools for advocacy, development, and the promotion of San culture and traditional knowledge. After the early stages, the Open Society Initiative of Southern Africa supported the further consultation and documentation of San values across borders. Job Morris has joined his father James in carrying on the project.

The Naro (Botswana) and =Khomani (South Africa) consultation processes have produced finished booklets now being used to inform NGOs and their work. The Ju|'hoansi (Namibia) booklet is almost completed, and the Khwe and four more language groups' booklets are still in process. As Kaelo Mokomo, the late director of KFO, wrote in his Foreword to the Naro booklet in 2012,

> [We] proudly present . . . a very special project of documenting the kind of values that people treasure and keep in spite of modernisation and development. This intensive and people-driven process of consulting, documenting and illustrating the leadership values is presented . . . in hope that they will also influence wider understanding in areas such as San education, representation and organisational processes. After comparing these perspectives with other San [Ncoakhoe] language groups in the region, a participatory process of conceptualisation of leadership development and identity building is planned so that the youth will not lose the beauty of their culture but also will not be left behind in a fast-moving new world. We hope that a community of practice will emerge from this process [to] not only serve the San but enrich the world on sustainable and legitimate systems of management and leadership. (Kuru Family of Organisations, 2012, 4)

As an anthropologist observing this process in a world today roiled with values and leadership issues, I want to underscore this strong statement. It says precisely why I do what I can to help preserve and promote San values I have been privileged to learn. In the =Khomani booklet, the intention is even more simply expressed: "(The group that met to list traditional values) felt that these values determine who they are as a people and how they live with each other and with other groups, but also how they live with the Land" (Kuru Family of Organisations, 2014, 2).

## !KHWA TTU SAN CULTURE AND EDUCATION CENTRE, TOURIST GUIDING PROGRAM, 2012–2017, YZERFONTEIN, SOUTH AFRICA

The legacy of apartheid has been more profoundly disruptive to San communities in South Africa than in the two countries further north where San people also live. Thus, an organization like !Khwa ttu has a difficult task in implementing consultative processes there. Nevertheless, calling on development workers, anthropologists, and community leaders with experience in Namibia and Botswana, this organization is now a vibrant joint venture between the San people, represented by the WIMSA, and the philanthropic Swiss foundation, Ubuntu. Through careful training in sustainable tourism and general life skills, !Khwa ttu supports San people in creating a safer, more self-determined future.

In particular, !Khwa ttu has for years specialized in the education and training of young San tourist guides, publishing in 2012 its comprehensive San Tourist Guiding Trainee Manual (!Khwa ttu: San Spirit Shared, n.d.). This Manual, of necessity geared to the local west coast environment north of Cape Town, South Africa, is so excellently structured that it can serve as a template for many such tourist guiding manuals for Indigenous groups all over the world. The reason it is so well structured is that it is based upon years of experience in a particular environment on a particular piece of land, with San students who feel strongly they are stewards of community knowledge that must be shared and saved.

The Guiding Manual provides a structured framework for !Khwa ttu's ecotourism docent course, but the broader program also encompasses life skills, computer skills, communication, entrepreneurship, work experience, and mentorship components. Each year, up to sixteen young San from all over southern Africa are selected to live at !Khwa ttu while completing the seven-month program. This plan has recently been augmented by three-month work internships for up to twenty-four San annually, in recognition of the economic realities faced today by individuals due to community fragmentation in the past. !Khwa ttu now offers opportunities not only in ecotour guiding but in other areas of tourism and hospitality, such as restaurant work, shop work, accommodation, vehicle and infrastructure maintenance, game farming, gardening, administration, and soon museum outreach.

Though each of the solutions discussed brings up contradictions, to me, each of the three examples well illustrates attempts at Erving Goffman's concept of "self-presentational energy" (Goffman, 1959). Each also presents a different, refreshing antidote to top-down models from the past.

## ACKNOWLEDGMENTS

I thank TUCSIN and Khwa ttu for the contributions they made to this chapter. I also thank the editors of this volume and Beatrice Sandelowsky and Chris Low for their suggestions on this chapter.

## REFERENCES

Biesele, Megan and Robert K. Hitchcock. *The Ju|'hoan San of Nyae Nyae and Namibian Independence: Development, Democracy, and Indigenous Voices in Southern Africa.* New York: Berghahn Booksm 2013.

Goffman, Erving. *The Presentation of Self in Everyday Life*. New York: Anchor, 1959.

Gosling, Melanie. *San Council L Code of Ethics for Researchers*. Cape Town: Ground Up, 2017.

Katz, Richard, Megan Biesele, and Verna St. Denis. *Healing Makes Our Hearts Happy: Spirituality and Cultural Transformation among the Kalahari Ju|'hoansi*. Rochester, Vt.: Inner Traditions International, 1997.

!Khwa ttu San Culture and Education Centre. *Tourist Guiding Programme, 2012-2017*, Yzerfontein, South Africa: !Khwa ttu San Culture and Education Centre, 2017.

!Khwa ttu San Culture and Education Centre. *San Spirit Shared .San Tourist Guiding: Nature and Cultural Guiding. Trainee Manual.* Yzerfontein, South Africa: !Khwa ttu San Culture and Education Centre, 2015.

Kuru Family of Organisations. *Naro San Values*. Letloa Trust: Ghanzi, Botswana. D'Kar, Botswana: Kuru Family of Organisations, 2012.

Kuru Family of Organisations. *The Way to Live: Values of the =Khomani San*. Letloa Trust: Ghanzi, Botswana. Kuru Family of Organisations, 2014.

Lee, Richard B. and Irven DeVore, eds. *Kalahari Hunter-Gatherers: Studies of the !Kung San and Their Neighbors*. Cambridge, MA: Harvard University Press, 1976.

Lee, Richard B. and Richard Daly, eds. *The Cambridge Encyclopedia of Hunters and Gatherers*. Cambridge, England: Cambridge University Press, 1999.

Nordling, Linda. San people of Africa draft code of ethics for researchers. *Science*, March 17, 2017. doi: 10.1126/science.aa10933.

South African San Institute. *San Code of Research Ethics*. Kimberley: SASI, 2017.

TUCSIN, The University Centre for Studies in Namibia, San public relations course 2002, in *TUCSIN 25 Years*. Windhoek: TUCSIN and WIMSA (Working Group of Indigenous Minorities in Southern Africa), 2004.

TUCSIN, The University Centre for Studies in Namibia.*Celebrating TUCSIN 35: Shaping the Future*. Windhoek: TUCSIN and WIMSA, 2013.

# *Chapter 5*

# "Your Soul Will Remember"

## *Reconnecting with the Kalahari, Celebrating Traditional Knowledge, Sharing the Meat, and Becoming a Community on the ǂKhomani San Lands Today*

R. Fleming Puckett

In the fields of human geography, cultural anthropology, tourism, and development studies, much excellent, important research has been conducted in recent years, highlighting both the positive potential of San-run tourism ventures (see, e.g., Giraudo 2018; Koot 2018) and the severe problems that can arise among San and other Indigenous peoples when tourist operations are imposed upon these communities by outsiders (see, e.g., Buntman 2002; Guenther 2002; Sylvain 2002, 2005). In this chapter, I will be profiling what is becoming a significant success story in the Kalahari today, produced by the dedication of several individuals and the confluence of multiple crucial factors.

I have argued (Puckett 2018) that only when the ǂKhomani San were able to organize themselves into their own "San-style" decision-making groups were they able to begin actively (re)engaging with their lands—and each other—after their successful 1999 land-claim settlement in South Africa's Kalahari Desert. The clearest evidence of this process in action is the success and defense of—and the income, community pride, ǂKhomani San identity, and traditional skills transfer generated by—the ǂKhomani San "Wildsplaas"[1] tourist and hunting ventures on Erin Farm.

Although these operations were initially contemplated by the ǂKhomani during their land-claim efforts in 1998 (Grossman, February 27, 2017), they were unable to develop or implement any of these plans during their first decade post-settlement. The Erin Farm hunt was launched as a pilot in 2012, but the renewed hope, the revived discussions, and the active planning stages were first clearly visible in 2009 and 2010, after the unofficial beginning of "administration" in 2008, when the last of the initial, unsuccessful ǂKhomani San Communal Property Association (CPA) Committees was dissolved by the South African government.

## BACKGROUND

Historically, the San people ranged across all of southern Africa, but encroachments by Bantu and white settlers led San territories to shrink to a small fraction of what they once were (see, e.g., Lee 1998a:5; Robins 2001a:7; Barnard 2007). The last of the remaining Indigenous San people of South Africa were evicted from their lands in waves (from 1931 to 1970) to create the Kalahari Gemsbok Park (now the Kgalagadi Transfrontier Park, or "the Park") and to keep it "pristine" so that tourists would not have to look at "hungry-looking Bushmen" (Guenther 2002:54).

When South African independence came in 1994, these southern Kalahari San people found themselves scattered across the landscape, working on white-owned sheep farms in the Northern Cape or performing for tourists at the Kagga Kamma Nature Reserve in the Western Cape (Chennells 2006). With independence came land reform, and in order to submit a valid claim for land "restitution" amid other competing claims, these individuals—many of whom had never actually met (Robins 2001a:26, 31)—had to form a target "community" to which land could be given under South Africa's Restitution of Land Rights Act, No. 22 of 1994 (the "Restitution Act").[2] With the help of attorney Dr. Roger Chennells, these San began to come together again, calling themselves the "ǂKhomani San" (Chennells 2006), and in 1999, the ǂKhomani San land claim was successful, granting this newly named (or renamed) "community" six farms—Erin, Witdraai, Miershoopan, Scotty's Fort, Andriesvale, and Uitkoms—outside the Park from which they had been evicted. The ǂKhomani land-claim settlement also included money for "development" and the purchase of additional land, which ultimately included Rolletjies and Sonderwater farms along the southern boundary of the Park. (See figure 5.1.)

Today, approximately 160 ǂKhomani San households live on the land-claim farms, and they range from four to ten individuals per household. These families, plus additional community members living in Askham, Welkom, Upington, and the surrounding areas, bring the total number of registered ǂKhomani community members to 1,400 in 2023. With many relatives identifying as San but not seeing the need for official "registration" as ǂKhomani

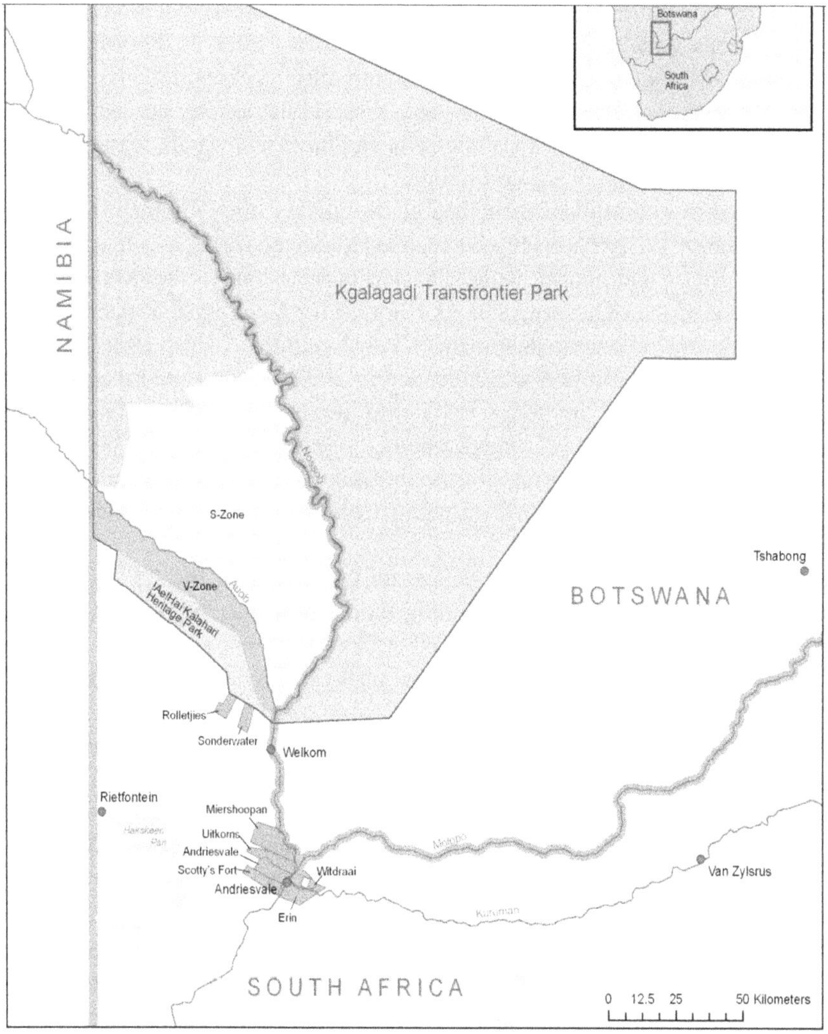


**Figure 5.1 ǂKhomani Land-Claim Area, Including all Eight Farms, the Park, and the Settlements of Welkom and Andriesvale.** *Source*: Map provided by Dr. David Grossman.

community members, the total ǂKhomani San population in South Africa today may actually exceed 3,000 (L. Steenkamp, January 28, 2023).

Along with the Restitution Act, this newly created or invented "community" (Chennells 2002:51; Sylvain 2002:1081–82; Kuper 2003:394; Guenther 2006:18) also had to comply with the CPA Act, No. 28 of 1996, or "CPA Act." In part, this Act defines "members" as members of a CPA or the members of a "community," which is itself defined as a group who "wishes to have its rights to or in particular property determined by *shared rules* under a written constitution" (CPA Act, §1; emphasis added). Such community constitutions tend to follow a standardized form, received by government officials—who seem to assume that a "community" will naturally have shared rules and values (Puckett 2018)—and these constitutions are adopted by communities in order to satisfy the laws and have their lands registered and recorded in the name of the community.

In the rush to complete their land claim quickly, the ǂKhomani Constitution of March 17, 1999, enshrined such standardized forms as a hierarchical, representative, community-wide decision-making body, the Management Committee of the ǂKhomani-San Association for Communal Property, or as it is commonly called, the ǂKhomani "CPA Committee." Even after the original ǂKhomani Constitution was amended in 2002, its mandated meeting and operating procedures for this CPA Committee still enshrined standardized concepts of "democracy," representation, organized voting, quorums, the taking of minutes, the hearing of formal resolutions, and the concept of—and reliance upon—"shared rules" (Second Constitution of the ǂKhomani-San CPA, adopted May 4, 2002).

Under this current (second) ǂKhomani Constitution, the CPA Committee has a variety of powers, including the authority to "manage the assets of the [CPA] on behalf of the members" and "to approve [and] compile . . . management plans for various land uses . . . including budgets, for the management and development of land" (ibid., §5). And even though this amended Constitution also established the new, smaller-group Ward Committees—including the Traditional Land Ward, which covers residents of and activities on Erin and Witdraai farms, and which was empowered to develop its own "standing rules with regard to its activities, composition, and election"—these committees were still entirely subject to the reporting and approval authority of the overarching, community-wide ǂKhomani CPA Committee (ibid., §6.3).

While such procedures and hierarchical structures might be quite familiar to white or Bantu agro-pastoral groups, the San would almost never have organized their societies in such a way, instead strongly preferring small-group organization, in which leaders emerge and decisions are made on the basis of consensus, adaptation, and fluidity (see, e.g., Lee 1998b:77, 79–84; Hitchcock and Biesele 2000:19–20; Barnard 2002:9, 2007:73), with for example, a group forming to undertake a particular task on a specific piece of land, and the leader chosen for his or her expertise in that task at that moment, but still possessing very little actual power or authority (see Smith et al. 2000:76). One ǂKhomani elder, the late Andries Steenkamp, commented on the CPA Act's insistence and reliance upon community hierarchies, arguing against the use of centralized, representative CPA Committees: "That style . . . is not a San style. That come from West Africa, not . . . southern Africa. [Having a hierarchy] is . . . not a San style" (A. Steenkamp, September 8, 2009). By pointing to West Africa, Andries refers to agro-pastoral governance systems developed by the Bantu-speaking peoples (including the Zulu), who appear, based primarily on linguistic studies, to have originated in West Africa (see, e.g., Oliver 1966).

For the ǂKhomani, the CPA Committee's centralized leadership and its management of land and assets "on behalf of" the ǂKhomani "community" were particularly problematic, as such structures have been for the San and other Indigenous groups in various regions. For example, speaking on behalf of San in Namibia, Kxao Moses ǂOma described the representative governance structures that were instituted in Namibia's Council of Traditional Leaders as follows: "We never wanted to represent our communities. That was a white people's idea in the first place" (quoted in /Useb 2001:24). Saugestad further explains that there is "a significant difference between the *participatory democracy* practiced by most hunter-gatherers, and the *representative democracy*, on which modern organizations are depending" (2001:315–16). Indeed, such "participatory," consensus-based decision-making would seem to require fluid, small-group organization, as large, static communities (and their representative committees) would never be able to reach the true consensus needed to make a final, lasting decision on any particular subject.

When such outside structures are imposed upon Indigenous groups—particularly those with an identity and history of hunting and gathering, like the ǂKhomani—the results tend to be severe personal strain (ibid.), as well as community apathy and inactivity, "failure" to operate the received structures, and eventually, Indigenous forms of governance emerging in a rather quiet, behind-the-scenes way over time (O'Malley 1996:313–15; see also von Bremen 2000:279).

In the case of the ǂKhomani, I have argued (Puckett 2018) that their ultimate lack of acceptance and understanding of the hierarchical, community-wide, representative decision-making structures, as received by government and enshrined in both iterations of the ǂKhomani Constitution, led to the failure of three successive ǂKhomani CPA Committees and several NGO-led "development" projects, plus the nearly complete lack of finalized land-use or development plans or visible socioeconomic progress (as reported by Chennells 2006; Bradstock 2005, 2006; Grant 2011) on the six ǂKhomani farms. (See figure 5.2.) Thus, although government labeled the ǂKhomani land-claim victory as the "rebirth" of their culture (quoted in Sylvain 2002:1081), the imposed decision-making and leadership structures severely impeded that "rebirth."

**Figure 5.2 An Example of a Failed "Development" Project: the ǂKhomani San "Sisen" Craft Shop.** *Source*: Photo by the author.

## OWNERSHIP OF ERIN FARM

Erin is a 15,000-acre farm that was part of the original ǂKhomani land-claim settlement, and with its rolling, grass-covered sand dunes, it seems endless. (See figure 5.3.)

Erin had previously been a profitable, white-owned sheep and game farm during apartheid, and after the ǂKhomani land-claim victory, the former owner had actually developed a scheme to have the farm returned to him. According to longtime ǂKhomani advocate and technical adviser Dr. David Grossman (January 20, 2010), the former Erin Farm owner began lending money to the newly resettled ǂKhomani residents, knowing that, with few—if any—sources of income at the time, the ǂKhomani would be unable to repay these debts. Former Department of Land Affairs (DLA)[3] representative for the ǂKhomani San, Peter Mokomele, also remembered this incident:

**Figure 5.3 The 6,000-hectare Erin Farm, the Most Southerly of the Six Original ǂKhomani Land-Claim Farms.** *Source*: Photo by the author.

> He did try. . . . What [the ǂKhomani] had done was that they had been buying from his shop and making debt at his shop and liquor store. So then they tried to pay with cheques, and the cheques would bounce. (15 September 2009)

When this occurred, as he had been expecting, the white farmer claimed Erin back as payment for these outstanding ǂKhomani debts. Dr. Grossman and his fellow technical adviser, Phillipa Holden Van Zyl, then brought in lawyers to try to resolve the dispute. However, a land auction was still scheduled, and on the day it was to proceed, the DLA agreed to use some of the ǂKhomani land-claim funds to pay off the ǂKhomani debts to the former Erin owner. As a result, Erin Farm remained in the possession of the ǂKhomani, but they were henceforth restricted from using any of their communally owned lands as security for debts (Grossman, January 20, 2010). Peter Mokomele summarizes, "That's why we're wary of them using the farm as collateral" (15 September 2009).

In fact, despite persuasive arguments against such restrictions (see, e.g., de Soto 2000), this lack of private ownership (i.e., individual title deeds) for land recipients is a common rule for many communities in South Africa today. Peter Mokomele explains as follows (also see figure 5.4):

**Figure 5.4 "RDP houses" in the ǂKhomani Settlement of Welkom, where Several ǂKhomani Families, including the Kruipers, Settled after their Eviction from the Park.** *Source*: Photo by the author.

> If you give a person a title deed now and he does not have anything, it's the same thing as if you buy someone who lives in a shack some farming implements and say, "OK, establish a small garden here. I will buy you seeds . . ." You *know* what he's going to do. . . . He has nothing—no means of income . . ., and now he has to work on an empty stomach? No. . . . He'd rather go and sell [the tools], fill the stomach at least. And now if you give someone a title deed, and he does not have anything, he's going to sell it, sell the house, which we have already seen with these RDP[4] houses [in other communities]. . . . Even when Government has put mechanisms in place that say, "You cannot sell this house," people have sold them. (2 September 2016)

Peter continued, with reference to Maslow (1943):

> So is it good for people to be getting title deeds? It depends on their circumstances. There are variables. If you have a base—if my uncle had left me a few cows, and if I have a house with a title deed, and I know if I sold it I could have money and I could do something, but I still have my cattle—then yes, it's a good thing. Because then I can do *other* things. It's a [form of] leverage. But like Maslow said, if I'm hungry, no. So it's a different matter until I'm satisfied;

then I can deal with higher level things. I can think. I can now innovate. I can now move, because I have the food and the shelter covered. I can now move to other things. The ǂKhomani San might not be ready. (2 September 2016)

Although the ǂKhomani did retain collective ownership of Erin Farm, they were able to do very little with it in the first decade of their ownership due to their difficulties in reaching land-use consensus (Puckett 2018).

Respected ǂKhomani community member and Witdraai Farm resident Jan Johannes "Oupa[5] Jan" Pietersen, referring to the then-living Traditional Leader and land-claim figurehead Dawid Kruiper, remembered as follows a decade after their historic victory:

Back in '99, the actual land claim, in our minds, was a *cultural* land claim. . . . Many people before that saw Oom Dawid Kruiper as Shaka Zulu,[6] but then the land claim came through, and there were farms and promises, and many people then moved back here [to the Kalahari], but they didn't think ahead of time, of what they were actually going to do here and how they were going to make it work. They were without plans. (1 September 2009; translated from Afrikaans by Gerbrand Nel)

And three successive ǂKhomani CPA Committees achieved no success in this sphere (Puckett 2018), struggling to manage community funds and failing to generate community-wide agreement on livelihoods, development priorities, or land-use plans.

## SUCCESS IN "FAILURE"

Under South Africa's land reform laws, when a land-recipient community proves itself incapable of operating the CPA structures effectively, the government can temporarily put that community under "administration," with a government employee or minister acting in lieu of the community's CPA Committee (CPA Act, §§11, 13, and 15). In cases in which a CPA "because of insolvency or maladministration or for any other cause is unwilling or unable to pay its debts or is unable to meet its obligations" or if the court determines it would be "just and equitable" to do so, the CPA may be liquidated and its land distributed by court order (ibid., §13). Thus, where a CPA Committee fails to perform its functions—making decisions, allocating parcels of land, and managing the community's lands overall—the options under the CPA Act include a new CPA Committee, "administration," or liquidation. However, aside from liquidation, a community's period of "administration" must come to an end with the election of a new CPA Committee, as there is no other organizational structure contemplated by the CPA Act. If a community falls under the CPA Act, it must learn to function with the CPA structure, as "communities" with "shared rules" managing commonly owned property in a democratic manner are key ideals espoused (and required) by South Africa's land reform program (see Puckett 2018).

Although done at first on an informal basis until their 2014 court case settlement,[7] the ǂKhomani were effectively put under "administration" in 2008, with the dissolution of the third ǂKhomani CPA Committee and the *de facto* appointment of DLA representative Peter Mokomele as final decision-maker for the community. It was under Peter's watchful eye that Indigenous forms of governance started to emerge behind the scenes, with the ǂKhomani, in the absence of a CPA Committee, beginning to "grasp the reins of power" in their own ways and "direct community activity in ways not intended" by government, as O'Malley (1996:313) had witnessed among Australian Aborigines who were facing similarly imposed structures. In the case of the ǂKhomani, this quiet, behind-the-scenes activity included the formation of smaller decision-making and governance bodies, like the ǂKhomani Stock-Farmers' Association and the Bushman Raad,[8] who concentrated on farms, families, and projects of interest to particular groups of ǂKhomani, rather than trying to force consensus across the entire community.

Under this new system, the Bushman Raad was particularly successful, as it focused on livelihood projects that involved "traditional" skills and tourism, the development of the Park and two farms (Erin and Witdraai), and the residents of one farm (Witdraai)[9] and one Park-adjacent settlement (Welkom). (See figure 5.5.)

To date, the most visibly fruitful of these projects has been the development of Erin Farm into a first-class game farm, tourism operation, and commercial hunt. Dr. Grossman recalled how the planning for Erin Farm began and then stalled:

> At a series of planning workshops [in 1998 and 1999]. . ., it was decided that Erin and Witdraai would be used for "traditional purposes." . . . Erin in particular was earmarked for game ranching, and these decisions were incorporated into the constitution. The first and subsequent CPA [Committees] totally neglected this decision, to the detriment of the more "traditional" people led by Dawid Kruiper. . . . [Before long] the game on Erin had been virtually wiped out. . . . So we now had the farm with damaged fences and no animals. (Grossman, 27 February 2017)

However, under both *unofficial* and then *official* "administration," the Bushman Raad was able to make decisions in "San-style" groups of ten to fifty individuals, rather than involving all farm residents or regional ǂKhomani

**Figure 5.5 Witdraai, One of the ǂKhomani "Traditional-use" Farms and Home of "Oupa Jan" Pietersen and the Original Bushman Raad Office.** *Source*: Photo by the author.

(see below). And these decisions could be developed into plans that could then be staffed, funded, organized, and implemented without the need to report to, seek approvals from, or funnel profits into a ǂKhomani CPA Committee.

As noted above and argued in Puckett (2018), this transformation began in 2008, when Peter Mokomele was made acting "administrator" for the ǂKhomani San. Indeed, during my initial fieldwork in the Kalahari in 2009 and 2010, the ǂKhomani—recently freed from the burden of the community-wide CPA Committee structures—were already throwing off the inactivity and frustrations of the prior decade and were beginning again to plan for the development of Erin Farm as a hunting operation.

For example, in January 2010, ǂKhomani elder Andries Steenkamp mentioned that David Grossman and Phillipa Holden Van Zyl had written a development plan for the ǂKhomani that included converting Erin into a game farm. The following month, a local white game farmer[10] was scheduled to visit Erin and help evaluate the condition of the land and fences "to help us make it ready" (January 23, 2010). Before Andries' untimely death on August

**Figure 5.6 Meat-Processing Training, Two Meat Refrigerators, and the Generators to Run Them Were All Part of the Infrastructure Needed to Launch Erin Farm.** *Source*: Photo by the author.

23, 2016, he had been both Chairman of the South African San Council and head of the ǂKhomani Stock-Farmers' Association and Sheep Bank. As the most actively involved ǂKhomani elder and a resident of Erin Farm, Andries would have been involved at a high level with the development of Erin. However, his clear focus on sheep farming meant that a larger proportion of the day-to-day Erin Farm details would have been handled by more "traditional" ǂKhomani,[11] particularly members of the Bushman Raad.

In the same month, the head of the Bushman Raad, "Oupa Jan" Pietersen, described their blossoming thinking and new activities related to Erin Farm at that time (also see figure 5.6):

> We are busy with the business plan, but the government gave Mier Municipality some money to help us with the fences and the slaughterhouse and all those things. The money is from the Department of Tourism. Our people had to go in

> for training to manage this—two from Erin, two from Witdraai, two from Miershoopan—I think it was 15 persons [total]. The training happened here in the Kalahari, but [Mr.] Du Plessis from GhoenasKraal[12] [company] in Pretoria gave the training. And he gave us certificates [approved by the Department of Tourism]. The training was on guiding, . . . hunting, [land and plant management], working with meat, game farming.[13] . . . This training was in October [2009]. All these trainings are finished. . . . By this training, we will be able to do all the jobs, practical and physical. We are ready to begin the job. (16 January 2010)

With pride, Oupa Jan showed me detailed maps that had been created for the traditional farms—at that time, Erin, Witdraai, and Miershoopan[14]—including the two new Park-adjacent farms, Sonderwater and Rolletjies, which had recently been procured with the additional land-purchase funding from their land-claim settlement. These maps, created in 2009 using handheld GPS devices with the assistance of University of Pretoria student Gerbrand Nel, showed all of the footpaths, roads, water points, medicinal plants, and various issues (including downed fences), of which there were 187 listed for Erin Farm alone. (See figure 5.7.)

**Figure 5.7 Turning Erin into a Navigable Game Farm for Hunters, Guides, and Tourists Required the ǂKhomani to Find, Map, Improve, and/or Create Roads in the Sand.** *Source*: Photo by the author.

Oupa Jan explained as follows:

> Community mapping through the cyber-trackers:[15] it's very basic community-orientated mapping. They've got the hand-held GPS machines with pre-set information, where you can actually just plot what you see. . . . That cyber-tracker information is then downloaded, and the community map is then updated. [Three cyber-trackers were out in the field that day on Erin and Witdraai farms.] The idea would be eventually to utilize all of the six cyber-tracker units on all of the farms . . . .as well as the . . . Park. (1 September 2009; translated from Afrikaans by Gerbrand Nel)

Indeed, in February 2010, they were planning to expand this project to begin documenting all of the ǂKhomani San "heritage places" (Pietersen, January 16, 2010). This expanded project was part of a campaign led by David Grossman and Phillipa Holden Van Zyl to have the "ǂKhomani Cultural Landscape" added to the list of "World Heritage Sites" by UNESCO. This campaign ultimately attained success in the summer of 2017 (see Cripps 2017).

During this time, there were still members of the ǂKhomani community grazing goats on Erin Farm. Thus, part of the process of creating the game farm included the ǂKhomani coming together in small groups to discuss their plans and their needs and to create solutions together. At the time, Oupa Jan described the contested land-use situation as "not OK," but he went on to

**Figure 5.8 ǂKhomani Livestock Alongside their Land-Claim Farms. The Process of Moving ǂKhomani Stock-Farmers into Designated "Sheep Camps," although not Easy, Demonstrated the Ability of the ǂKhomani to Discuss, Negotiate, and Make Decisions in their Own Ways.** *Source*: Photo by the author.

add, "we are busy to make another camp[16] on Miershoopan to share it, take these farmers to this place, then we will carry on. In February we will make those camps. Next month, we will make the fence right" (January 16, 2010). (See figure 5.8.)

Thus, in 2009 and 2010, the ǂKhomani were able to come together—in the absence of a CPA Committee—and negotiate in their own way to create solutions that today are beginning, directly or indirectly, to benefit the entire community, but which at the time did not involve an imposed, community-wide committee structure.

Peter Mokomele, as acting DLA "administrator" for the ǂKhomani from 2008 to 2013, remembers this process, in which his efforts ultimately helped to replace the imposed former committee structure with discussions, negotiations, and forms of decision-making that came from the ǂKhomani themselves:

> When I started working with Andries [Steenkamp] on the camps, . . . there was game on Uitkoms and Miershoopan. So I said, "Guys, let's sit. Let's talk." So for Andries, the first task I gave him was, "Let's determine who has livestock and where that livestock is." So we had an audit. . . . [Andries and his team] came back and gave me a report and said, "These are the people who have livestock." So then we agreed, "OK, we're going to make camps there. We're going to leave Erin."
>
> We developed the plans in small steps. So now . . . we had to . . . persuade the farmers that were in Erin to get out, so that now more game can move in there. . . . So it was a process of negotiations, but . . . as long as I am from outside, I was not going to be saying to them, "Tomorrow, you are going to be doing 1, 2, 3, 4." I was going there and saying, "OK, let's sit down, and you tell me what you think is best. So let's plan, but *I'm not going to be telling you how to plan. You are going to tell me how we should be planning*." So they told me. Then we applied whatever they were saying. So that's why they could agree. . . . No one could come there and bulldoze them. . . .
>
> For me, . . . it was more helping them to define themselves and therefore to work out their own plans, to say, "*If we find our own arrangements, we can then plan*". Like with the Aborigines or other groups, as long as they have been in the land for a longer period—they have to *find the rules of engagement*—their cultures and their customs become stronger, their traditions. They . . . interact better with outsiders because they have defined themselves. . . . I allowed that flexibility. But it was also more . . .. We had meetings every second week. I went there. *They said . . . they didn't want a committee*,. . . so that, if a decision is taken, it includes them, and we can negotiate whatever needs to be negotiated. So that was the recipe. The recipe was to say, "If I am coming there, we are going to have a 'forum' meeting," as we called it. So *it did not become a problem of a small committee which then...had to make decisions for the whole community*. . . . The meetings that we had bi-weekly were to open the process. (2 September 2016; emphasis added)

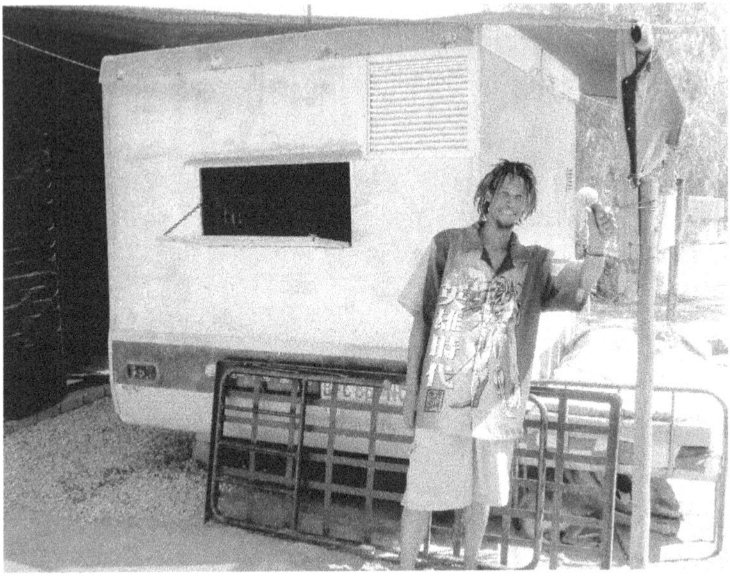

**Figure 5.9 Jan Johannes "Oupa Jan" Pietersen, Former Head of the Bushman Raad, is, as Dawid Kruiper had been, a Visionary San-style Expert/Leader who was Instrumental in Erin Farm's Early Development.** *Source*: Photo by the author.

With this more open, "San-style" process replacing the forced community-wide structures, ǂKhomani collaboration actually (and for Government, probably *surprisingly*) increased, and new agreements, ideas, plans, and real enthusiasm began to emerge on the ǂKhomani lands for the first time since 1999 (Puckett 2018). (See figure 5.9.)

In both 2009 and 2010, Oupa Jan expressed clear exhilaration regarding their rekindled plans for Erin while under "administration" (also see figure 5.10):

> Erin could be a very good farm for wildlife, where we have ideas for hunting, the working of skins, taxidermy, game drives, four-by-four routes, and commercial hunting. The idea is that Erin could work as a commercial wildlife farm that way; hunting could [one day] take place on a larger scale to produce biltong[17]. . . to be sold. It's definitely got a very strong business motivation. . . . It must act as an economically viable center. One of the first steps, from April this year, was to start up and look for funding for our own office to manage, amongst other things, Erin's administrative side. That has taken place (1 September 2009; translated from Afrikaans by Gerbrand Nel).
>
> And in April [2010], Tswalu [Kalahari Game Reserve (see Endnote 10)] will give us wild game, and then we will carry on with our business plan. We will make 4-wheel-drive trails, you can look for the wildlife, you can guide, and all

> these things. . . . [Mr. Oppenheimer] is a farmer here with a wild game park. He will sponsor us: not money, but springbok, gemsbok, and blue wildebeest. . . . People will come from England and other foreign countries to our country, and they will like to see the Bushmen. We will [develop the farm] for three years. . . .[18] It's a partnership. (16 January 2010)

Oupa Jan's comments regarding their work to develop a business plan, obtain funding, create partnerships, attract overseas tourists, and administer Erin Farm themselves "as an economically viable center" stand in sharp contrast to the belief expressed during my fieldwork by many outsiders—including several of their non-San neighbors in the Kalahari: government representatives, businessmen, tour operators, Park employees, and even schoolteachers—that the ǂKhomani are either unwilling or unable to manage business ventures. Yes, the ǂKhomani did receive, and continue to seek, business, financial, and computer training, and yes, they did receive enormous assistance from David Grossman and Phillipa Holden Van Zyl during this period, but the fact that Erin Farm has continued to prosper in the absence of David, Phillipa, and Peter Mokomele is testament to the ability—and adaptability[19]—of the ǂKhomani themselves. For their first decade back on the land, the ǂKhomani were hindered by their inability to reach consensus and inspire

**Figure 5.10 A Herd of Gemsbok, One of the Species Donated to the ǂKhomani by the Peace Parks Foundation, South African National Parks, and Tswalu Kalahari Game Reserve.** *Source*: Photo by the author.

activity under an unfamiliar set of imposed "community" structures, but *not* by any inherent inability to lead, motivate, collaborate, and run business ventures according to their own cultural norms as, in Oupa Jan's terms, "the Bushmen."

Thus, as the ǂKhomani began to flourish under their own re-emerging small-group structures, they eagerly sought to learn—from outsiders and from each other—as much as they could to help them adapt to their new circumstances. Vinkie van der Westhuizen, daughter of the late ǂKhomani Traditional Healer, "Oom Jan," explained as follows in 2009:

> Now I am just busy with tourism, helping with the guides and trackers to do trainings—tracker trainings—and for the guides, I do the general history, the heritage sites, and how to manage with tourists. I'm still learning. I'm learning every day. ...We do animal tracking, bird tracking, traditional plants. . . . I was a local guide, but I went to a training for "National Tour Guide." So I am a national tour guide and also a bird guide now.
>
> So I am busy to [teach other ǂKhomani] all the things I was learning in my training. And I am still learning, to help them to be comfortable with tourists. . . . Most of our people can't speak English. So all of us are struggling, but we are still busy learning. . . . I want to help them speak English—teach them. So now we are busy to [have] guide and tracker meetings every Thursday. So that we can talk together and do the general history, how to go out with tourists, how to act with them.
>
> Northern Cape Tourism is also helping the guides with trainings, . . . with marketing. . . . There is other funding from Ford Foundation, the African Safari Lodge Foundation,[20] [for] . . . the cyber-tracker system. And there is also funding for...the guide training. In that training, the guides were [learning about] nature, the culture, rifles, and first aid. (8 September 2009)

Here, in addition to the importance of outside guidance and funding—procured initially by David and Phillipa—there is evidence of traditional San structures. Those ǂKhomani who were interested in pursuing livelihoods based on tourism, guiding, tracking, and hunting on the ǂKhomani farms were coming together as a small group to meet, discuss, share, and learn from each other, rather than feeling any compulsion to include larger groups or dissenting ǂKhomani, of whom there were many (see below). Typical San adaptability is also evident here, as these ǂKhomani were clearly seeking to learn the full breadth of subjects—cultural, natural, practical, professional, behavioral—needed to run successful tourism and hunting ventures on Witdraai and Erin farms. (See figure 5.11.)

**Figure 5.11 A Meeting of the ǂKhomani Tracker-Training Program Being Held on Miershoopan Farm.** *Source*: Photo by the author.

## BENEFITS TO THE ǂKHOMANI

The ǂKhomani now have a secure, state-of-the-art fence around Erin Farm, tented "safari-style" tourist camps with hot and cold water, a skinning and meat-storage station that includes walk-in refrigerators attached to generators, a shooting range for sighting-in clients' rifles, a number of African species that include gemsbok (*Oryx gazella*), springbok, red hartebeest, eland, and blue wildebeest for hunting, and over the years, various non-hunted species, including ostriches, steenbok, giraffes, zebra, a single impala, and even camels. (See figure 5.12.)

Erin Farm has impressive roadside signs with the "Your Soul Will Remember" slogan, and the employees wear "Erin Wildsplaas" uniforms with great pride. Erin's hunting operations—which, as noted above, began with a small pilot season in 2012—are now fully booked year over year by both domestic and international hunting parties, and there are plans to increase non-hunting tourist activities on Erin during the off season, including four-by-four drives, guided tracking, birdwatching, and meerkat photo shoots.[21] (See figure 5.13.)

**Figure 5.12 Camels on Erin Farm. Not All of the Species on Erin Are Hunted**! *Source*: Photo by the author.

**Figure 5.13 Impressive Roadside Signs—Set along the Highway that Tourists would Take either Westward to Namibia or Northward to the Kgalagadi Transfrontier Park—Showcase Erin Farm Activities.** *Source*: Photo by the author.

Although tourism is the most popular concentration for ǂKhomani students who attend high school, there are actually few tourism opportunities in the Northern Cape today. Erin Farm is providing such an opportunity, as ǂKhomani can serve as tour guides and translators, as well as help in the tourist camps. In addition, as found by Wenzel (2016) among the Inuit, Erin provides employment opportunities for individuals with important traditional skills that might not be valued in the general marketplace, such as tracking, the preparation of hides, and the processing of meat. (See figure 5.14.) However, unlike in the Arctic, the hunting clientele in South Africa shows no signs of becoming "fickle" (ibid.:56).

Beyond employment for six permanent game rangers, two to three office managers, and several seasonal guides, trackers, drivers, and skinners, Erin is producing great community pride as a premier destination, and those involved in its operation are earning money and are the objects of community admiration, as well as occasional envy. (See figure 5.15.)

These employment opportunities and community pride are, in turn, serving to validate San traditional ecological knowledge among the (re)created "ǂKhomani San" land-claim "community." In September 2009 (well before

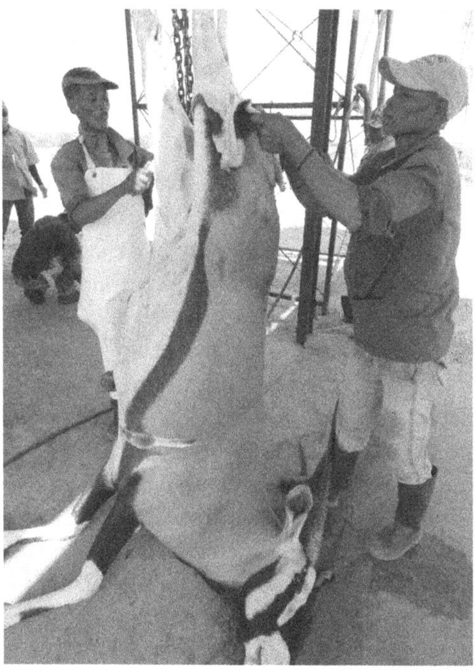

**Figure 5.14 Skinners Jan Matthys and Koos Brou Demonstrate, Teach, and Earn Money Practicing their Traditional Skills on Erin Farm.** *Source*: Photo by the author.

**Figure 5.15 As a Hunting Guide and Tracker on Erin Farm, Johannes Tieties has Gained Income, Shared Big-game Meat, Taught Younger ǂKhomani, and Earned great Respect in the Community. As of 2024, Johannes is a Member of the ǂKhomani Committee of Elders and has also Launched a new Youth Sports Program Funded by the Kalahari Peoples Fund (KPF).** *Source*: Photo by the author.

any Erin Farm ventures were launched), Oupa Jan Pietersen was looking ahead, expressing his belief that, properly funded, a game farm would be both "economically viable" and a place "where we can also practice our traditions." He also noted that the ǂKhomani tracker-training program mentioned above by Vinkie was originally organized on Miershoopan Farm by the South African San Institute NGO (SASI), but was being run by the Bushman Raad, who considered it a project to benefit the whole ǂKhomani community. Perhaps because of the importance of tracking as a central San skill and an ability that is sought after by both tourists and hunters, Oupa Jan stated that this "is one of the projects that at least do not have signs of dying out or getting killed" (September 1, 2009). During a time when SASI was being forced to downsize due to funding shortages, leading to offices closing and development projects being abandoned by SASI, this is a doubly powerful statement that, in a sense, the tracker-training project was too important or central to the ǂKhomani people, their identities as San, and their future plans to allow it to "die." Sadly, funding and personnel losses did ultimately bring about the demise of SASI, as reported by Roger Chennells in 2020: "SASI is basically all but moribund, which brings to a virtual end an NGO that tried hard over much of two decades to provide some stability to the San" (Chennells 2020).

Thankfully, many ǂKhomani have also worked hard to ensure that traditional skill-transfer opportunities have continued on their Kalahari lands. On Erin Farm, for example, the ǂKhomani have introduced a new "pairing" scheme, in which younger ǂKhomani join experienced elders in the field during hunts, as described by one of the primary Erin Farm administrative managers, Luce Steenkamp:

> The reason for using older people for the skinning...is because they have years of experience with . . . trophy animals. The younger ones we also use for skinning . . ., but [they] need more practice to do it perfectly. There are programmes in place . . . to transfer the skills from the elders to [the] younger generation; they just need to practice regularly . . . . There was even last year a refresher training for the younger skinners and guides and we usually try to mix them, to work together with the elders. (1 March 2017)

Indeed, similar impacts—revalorizing "the bush" and perpetuating traditional knowledge—were noted by Lee (2016) with respect to the commercial hunting operations in Namibia's Nyae Nyae Conservancy, to which the Erin founders looked for comparison and inspiration (Grossman, March 1, 2017).

Part of this revival of interest in those traditional San skills that were undervalued post-eviction (and were therefore lost by many ǂKhomani during the twentieth century) includes seeking specialized training, not only from ǂKhomani elders, but also from outsiders, as noted above by Vinkie van der Westhuizen. Dr. Grossman, for example, mentioned the importance of training by GhoenasKraal, which in addition to tour-guide training, has also included hunt-specific training in carcass preparation, meat processing, and cape preparation (i.e., trophy shoulder-mount taxidermy) (March 1, 2017).

But in this hunting operation, the animals taken from Erin Farm do not ultimately benefit the hunters alone. The meat remains the property of the ǂKhomani. And although Hitchcock & Sapignoli (2016) found shrinking sharing networks among San in the Central Kalahari of Botswana, the ǂKhomani are working to ensure that parcels of hunted meat are reaching every member of the community living on the farms. This ties closely to Woodburn's (1982) observations among the Hadza, where large-animal meat sharing is essentially an obligation imposed by the community, and I witnessed some indication of pride among the Erin workers in being able to help fulfill this obligation on the ǂKhomani lands. (See figure 5.16.)

Ultimately, the Erin Farm "Wildsplaas" is perhaps the best-run San-led commercial hunting operation in southern Africa. This project—planned with care and operated with dedication and pride—has increased the use, prestige,

**Figure 5.16 Johannes Tieties Prepares to Share Gemsbok Meat with Members of the Erin Farm Staff. One of the Operations Managers of Erin, Frederick Padmaker, also Ensures that as much Meat as Possible is Shared among the ‡Khomani Farm Residents.** *Source*: Photo by the author.

**Figure 5.17 Barbara and Arie Raats at their Living Museum Exhibit, across from the Main Entrance to Erin Farm. They are beginning to Benefit from the additional Tourists Brought to the Area by Erin, the Park, and the New "World Heritage" Status of the ‡Khomani Lands.** *Source*: Photo by the author.

and transfer of traditional knowledge, provided meat to community members, generated significant earnings (see below), and either directly (through profits added initially to Bushman Raad accounts) or indirectly (by drawing new tourists to the area) fed other ǂKhomani tourism ventures. These currently include Aunt Koera's Kalahari Kitchen, run by the wife of the late Andries Steenkamp, and the Living Museum and Boesmanrus grass-hut accommodations, both operated by Barbara & Arie Raats. (See figure 5.17.)

## REASONS FOR SUCCESS

To summarize, the success of Erin Farm as a tourism operation and its positive impacts on the ǂKhomani community appear to have been the result of several important factors.

First, outside technical advisers Phillipa Holden Van Zyl and David Grossman worked tirelessly with Traditional Leader Dawid Kruiper, the DLA's Peter Mokomele, and the head of the ǂKhomani Stock-Farmers' Association, the late Andries Steenkamp, to develop land-use and camp-allocation plans, apply for funding, and negotiate with the stock-farming ǂKhomani, who would have to be relocated from Erin to other ǂKhomani farms.

Second, key donations of game animals by the Tswalu Kalahari Game Reserve and South African National Parks, funding via the Peace Parks Foundation (in part from Rotary and BMZ, the German Federal Ministry of Economic Cooperation and Development), a low-interest loan (20 years at 2 percent) from JAMMA International,[22] DLA development funds, and training paid for by the South African Lottery were all essential in getting Erin off the ground operationally, not only in terms of staff education but also on the infrastructure side, including the purchase, capture, and transportation of game animals to Erin, secure fencing, water-source improvements, and the tented camp for visitors to the farm. And today, commercial farmers from Wildlife Ranching SA are helping the ǂKhomani fulfill the important tasks of counting their game animals and setting hunting quotas for each season (Grossman, February 27, 2017).

Next, the dedication of the two initial ǂKhomani managers of Erin Farm, Luce Steenkamp and Frederick Padmaker, was also critically important, particularly as they persevered in a clearly demanding job for quite some time without adequate support staff. Luce explained,

> For the last couple of years, I have spent most of my working hours on Erin administration to make it work, say 5 hours out of the 8 hours. I think even Fred, who has also had other responsibilities, also had to spend more of the time on Erin but not on administration but operational staff and keeping the hunters or guests happy. For this year Erin hunting was not the same, as Fred has been to college for the year. I have given the hunting bookings over to my colleague

> and [am] only managing the administration and finance side of Erin, as [I] am having more responsibilities. (L. Steenkamp, 15 November 2017)

This colleague is Dirk Pienaar, now the ǂKhomani San Tourism Officer (Pienaar, December 4, 2017), who, along with Frederick Padmaker, has received a diploma in Wildlife Management (Grossman, February 27, 2017). As of 2024, Dirk is heavily involved in helping to manage Erin Farm operations, bookings, and administration. Once Frederick returned from his course of study, Luce was temporarily able to depart the Kalahari to undertake her own further studies, although not to the extent that she had hoped (L. Steenkamp, December 7, 2020).

In addition to the hard work of these ǂKhomani individuals, a very skilled professional hunter (PH), Le Roux Coetzee,[23] based outside Kathu in the nearby Green Kalahari, has effectively volunteered his time, accepting reduced daily rates to assist the ǂKhomani in meeting the legal requirement of having a licensed PH on staff to guide overseas hunters. (See figure 5.18.)

**Figure 5.18 PH Le Roux Coetzee (Left) Guides an Overseas Hunter (Center) on Erin Farm. The Non-San Tracker and Gun-bearer (Right) Stayed in Barbara and Arie Raats's Grass-hut Accommodations during the Week-long Hunt**. *Source*: Photo by the author.

According to Dr. Grossman (February 27, 2017) and Dirk Pienaar (June 30, 2023), the ǂKhomani are hoping for one of their own community members to become qualified as a South African PH someday, as attracting and guiding overseas hunters—who are charged much higher daily rates and trophy fees—is a particularly lucrative segment of today's South African economy.

Finally, I argue that the launch and initial success of Erin Farm could not have come about during the prior period of ǂKhomani community-wide CPA Committees, from 1999 to 2008. Ideas for creating a ǂKhomani game farm, first discussed among these San land claimants in 1998 and 1999, remained dormant until the third CPA Committee was disbanded. In fact, a game farm was not a priority for many community members, who see tourism as neither sustainable nor reliable in the dry, "remote" Kalahari, with Erin Farm lying more than nine hours' drive from the closest international airport, O. R. Tambo in Johannesburg, and approximately eleven hours from Cape Town International Airport.

The success of Erin Farm required focused, small-group decision-making among those ǂKhomani who believed that an Erin tourism operation could succeed. During "administration"—and with Peter Mokomele's unwavering support, assistance, and personal commitment—the Bushman Raad could make consensus-based decisions involving "traditional" San skills and livelihoods, their plans could proceed without the need for the approval of a large, overarching committee, and passionate experts like Oupa Jan, Vinkie, Johannes, Luce, Frederick, and Dawid Kruiper's grandson "Klein Dawid" could emerge to take on specific roles and tasks as they arose. (See figure 5.19.)

David Grossman recalls that, early on, the Erin/Witdraai Ward and the Kruiper family "co-operated fully" as they held monthly meetings to discuss plans and management ideas, and it was this group who nominated particular ǂKhomani San to become Erin Farm employees (February 27, 2017). In typical San style, those actively involved in daily discussions and decision-making about ideas, roles, and plans for Erin ranged, as noted above, from ten to fifty ǂKhomani, depending on the magnitude of the decision being considered (Grossman, March 1, 2017). This is significant, given that the size of the ǂKhomani community at the time was, according to Andries Steenkamp, 1,024 members, including 360 living on the farms and 664 living in various towns across the Northern Cape (January 23, 2010; see also Appendix 1). Prior to "administration," CPA Committee decisions would have had to represent the interests of—and accommodate ideas, input, and criticisms from—of all of these ǂKhomani community members (Second ǂKhomani Constitution, §5).

**Figure 5.19 Operations Manager Frederick Padmaker (Right) Working with the Grandson of Dawid Kruiper, called "Klein (or Small) Dawid," on the Erin Shooting/Practice Range.** *Source*: Photo by the author.

"Oupa Jan" Pietersen—a member of the ǂKhomani "conservation area" of Witdraai, Erin, and the "contract" section of the Park[24]—described the impact of the collapse of this CPA Committee system:

> In the beginning, there was a CPA [Committee] that represented the interests of the [whole] community. . . . Then, . . . people realized how many of the community were also involved in stock-farming, and then there was an over-emphasis on to the stock-farming side. Investment moved away from the cultural side of things into the stock-farming. And there was mismanagement. The CPA [Committee] failed the community. None of the expected goals were reached, and this resulted in the CPA [Committee] being nullified.
>
> After 2007, we decided then, *because of the failure of the CPA [Committee]* and also their lack of interest in the conservation side of the people's interests, we decided to actually return to the original plan, which the people were speaking about in '99, and this rather includes looking more closely at [commercial hunting and] the conservation side of things. (September 1, 2009; translated from Afrikaans by Gerbrand Nel; emphasis added)

Thus, the primary reason for the decision by those ǂKhomani who wanted to pursue more "traditional" San livelihood options to revive their 1999 discussions about a ǂKhomani hunting venture was the beginning of the unofficial "administration" period, which began after the failure of the third ǂKhomani CPA Committee.

As consensus-based decision-makers, these San were finally freed from the need to reach such consensus with the entire ǂKhomani population. And this was particularly important to these "traditional"-livelihood San, who faced significant criticism from most of their stock-farming compatriots:

> Many [ǂKhomani] do understand [nature and conservation], but the lack of knowledge or the fact that many didn't grow up in the area or close enough to nature, that leads them to not understand it fully . . ., and it leads them to throw stones at others who feel more strongly about it. Three out of three of the people—a large portion—are often critics, like to be critical of the people that actually try and make things work. (Pietersen, 1 September 2009; translated from Afrikaans by Gerbrand Nel)

As will be seen below, teasing, criticism, and banter are typical of San societies, but the period of ǂKhomani "administration"—"unofficial" from 2008 to 2013 and "official" from 2014 to mid-2021 (see Endnote 7), when the DLA ended Terance Fife's contract without appointing a replacement "administrator" (L. Steenkamp, March 17, 2021, and June 16, 2021)—has made room for separate livelihoods and different ideas on how to make a living in the Kalahari. Many of the criticisms specifically regarding Erin Farm continue today (Grossman, February 27, 2017), but starting in 2008, the ǂKhomani became free to fragment in typical San fashion and create fluid, small-group structures to lead individual livelihood projects. Indeed, even Andries Steenkamp, who was so instrumental in working with Peter Mokomele to establish Erin as a game farm free from sheep, goats, and gaps in the fences, did not agree that game farming was the answer to a sustainable future for the ǂKhomani people:

> The previous owner of Erin only got into game farming after he was a successful sheep farmer—and *only* to make extra money! His main income was from sheep. That is the Kalahari! You can bring tourism—OK. I promise you, you can never make money in tourism. The San shall never make good money with these farms in tourism. It is only land to do stock-farming! (23 January 2010)

Still, separately conceived livelihood projects and small-group leadership structures on the ǂKhomani lands allow this typical San dissent and criticism to exist, while not jeopardizing the work of other community members. Interestingly, at Andries' funeral on September 3, 2016, the meat to feed the large crowd of mourners from across South Africa was provided by overseas tourists from their hunt on Erin Farm.

## THREATS TO CONTINUED SUCCESS

Thus, external funding, outside guidance, ǂKhomani dedication, and room for "San-style" discussion, decision-making, and separate livelihoods were all critical in providing the spark and momentum needed for Erin Farm to become a going concern. There are, however, a few potential threats to the future of Erin Farm as a sustainable tourism venture.

First, while "[o]nly half of Africans have heard of climate change" (*The Economist*, August 19, 2023) and predictions on the impacts of climate change on the Kalahari ecosystem specifically have varied widely, the current yearly trend is toward increased drought. If this continues, it will significantly impact this community of San people, who—even twenty-five years after their land-claim victory—still do not have reliable water sources for themselves, their livestock, or wild game. Initially, there was evidence that the Erin Farm game were reproducing well, with one 2016 visitor, for example, reporting large numbers of springbok fawns at that time (R. L. Puckett, January 11, 2018). During that period, however, ǂKhomani Tourism Officer Dirk Pienaar also stated that Erin Farm had been operating on a year-to-year basis, with some doubts each year as to "what is available to hunt for the next year" (December 4, 2017). Thus on Erin, careful, sustainable game management—including counts and quotas—will need to be maintained. Dr. Grossman agreed that the "increasing frequency of droughts affecting animal numbers" would be a significant threat (February 27, 2017), and sadly, 2019 in particular saw significant numbers of wild game on both Erin and Miershoopan Farms dying of starvation due to severe drought conditions in the Northern Cape.

Second, Erin must remain profitable in order to continue making its loan payments to JAMMA International, paying staff salaries after the Rotary grant is exhausted, and saving small amounts of money for a rainy—or very dry—day. (See figure 5.20.) With long-time technical adviser David Grossman having passed away after retiring to South Africa's Eastern Cape and Phillipa Holden Van Zyl now living in Australia, there is also some concern that future grants will be fewer and further between, as David and Phillipa prepared and signed all (or most) of the ǂKhomani funding applications in the past. Luce Steenkamp expressed similar concerns about the new CPA Committee (see below), who are now operating without the official oversight of a government-appointed "administrator":

> Regarding the CPA Committee, they are . . . trying their best and do what they can to keep things going. Although . . . I feel that there is still a need for someone from outside to manage the assets [of the ǂKhomani San] even if it is for another 4 years. By then someone from the community might be able to manage things. . . . [We also] need some intervention to train those who are interested to stand for elections; otherwise, we will still have the problem of committees not understanding their roles. (21 February 2022)

**Figure 5.20 In Addition to the Need to Continue Repaying the initial "Wildsplaas" Startup Loans, the Continued Success of Erin will Depend on the Ability to Manage Hunting Quotas, Maintain Facilities, and Pay Salaries to Managers, Staff, and Seasonal Workers.** *Source*: Photo by the author.

Not directly impacting Erin Farm, but a troubling finance-related development nonetheless, there is recent evidence that this current ǂKhomani CPA Committee has failed in at least some of its bookkeeping and reporting duties. The initial result is an absence of funding from the Department of Social Development for the ǂKhomani San preschool, leading to a lack of teacher salaries and daily lunches for thirty toddlers (Pienaar, October 20, 2023 and November 3, 2023).

According to Dr. Grossman (February 27, 2017), Erin generated 800,000 Rand in 2016, which equates to approximately US$43,400 in November 2023. Loan payment amounts have been 160,000 Rand (around US$8,700 at November 2023 exchange rates), "all generated from hunting," and as noted above, current ǂKhomani plans include training their own community PH "to increase foreign hunters," who are charged higher rates (ibid.). However, despite David's prior reports that, over time, "bookings have increased" and "some [hunters] have become regulars," it seems that, in order to help mitigate the risk of reduced game available for hunting—through slower-than-expected breeding, over-hunting, or less reliable food and water sources due to ongoing drought—the ǂKhomani will need to proceed with those plans mentioned above (including meerkat photo tours) that will help diversify their income sources and increase their non-hunting-season revenues, since South African tourism has picked up again post-pandemic.

In the area of traditional skills, there is, of course, the threat that younger generations will not value the skills of their elders or that those young men and women who have gained such skills through Erin and the *veldskool*[25] will leave the ǂKhomani farms to earn more money elsewhere. Indeed, this has started to happen (Grossman, February 27, 2017). And with the deaths of Dawid Kruiper (June 13, 2012) and Andries Steenkamp (August 23, 2016), community members began to express concern that the new progress and emerging "community" identity and pride might subside in the absence of new "traditional" and stock-farming experts, advisers, and leaders. Peter Mokomele, in talking about the inevitable end of ǂKhomani "administration," had mentioned this issue:

> Government has to pay an administrator. But also the person doesn't stay there for very long. . . . You see, people like Andries [Steenkamp]: you'd have thought that, if they were still alive, you could devolve power to them. And say, "OK, Andries, you take over as the administrator. You come from here. You are a San. So go on with it!" But now, I don't know who else will be left there, who one could identify. It was a big blow when we heard that he is gone. (2 September 2016)

Two years later, on April 7, 2018, Petrus Vaalbooi was inaugurated by the ǂKhomani San as their new Traditional Leader.[26] Although originally serving as the first leader of the group of ǂKhomani who prefer stock-farming livelihoods (and a member of the most prominent ǂKhomani stock-farming family), Petrus still seems to express a strong connection to a "Bushman" identity (see Puckett 2018). However, his dedication to, and work for, the broader ǂKhomani community to date have not begun to approach that of prior community elders, especially Dawid Kruiper and Andries Steenkamp. Thus, it remains to be seen whether the previous ǂKhomani respect for Dawid as a traditional *veldskool* mentor or Peter's belief in Andries as a natural candidate for "administrator" or farm manager might one day apply to Petrus as well.

In fact, it was reported in October 2020 that Petrus Vaalbooi led an illegal hunt on Miershoopan Farm, where his group—including a hired hunter from outside the community (paid in game meat)—allegedly killed over 100 springbok, despite protests from many ǂKhomani community members, who were forced to call upon the Stock Theft Unit of the South African Police to intervene—against their own Traditional Leader—before the hunting group reached Erin Farm, where they intended to continue the killing (Fife, October

12, 2020; L. Steenkamp, October 16, 2020). Erin's manager, Luce Steenkamp, described her feelings after the incident:

> It is clear that most members are not in support of what the traditional leader and his team did . . . . Right now, we're also not sure how we will get back what the [ǂKhomani] have lost . . . [including] trust. (October 16, 2020)

Amid these sad developments, it is also clear, as Peter noted above, that the lengthy and expensive period of ǂKhomani "administration" by the DLA/DALRRD had to come to an end. Although Terance Fife's government funding for his role as official administrator for the ǂKhomani (see Endnote 7) ended on March 31, 2021, a new ǂKhomani CPA Committee had, in fact, been elected back in 2017. This new Committee has the former Secretary (under Andries Steenkamp) of the South African San Council, Collin Louw, as its Chairperson, and once administration was set to end with an official handover by Terance to the ǂKhomani community in mid-April 2021 (Fife, March 25, 2021), this body once again took on the duty of acting on its own authority and attempting to govern the ǂKhomani farms on behalf of the community as a whole for the first time since 2008. In addition, the formerly independent Bushman Raad and ǂKhomani Stock-Farmers' Association have been absorbed into the new CPA Committee (Pienaar, August 30, 2019), falling under the ǂKhomani Constitution's current Traditional Land Ward[27] and Stock-Farms Ward structures, respectively.

However, with the return of a fully empowered, overarching ǂKhomani CPA Committee acting without an administrator's checks and approvals, it is possible that the typical San criticisms, jealousies, and banter—elevated to, and given the authority of, a centralized, community-wide "pulpit"—will jeopardize the ability of the former Bushman Raad (now Traditional Land Ward Committee) to focus its energies, funding, and human resources on the perpetuation of Erin Farm, the planned development of Miershoopan Farm, and the training and knowledge-transfer activities associated with these farms and the Park. Indeed, three years before the illegal hunting incident described above, Luce Steenkamp found that jealousy, envy, and lack were fuelling criticisms against Erin Farm by other ǂKhomani San:

> People are proud of Erin but the co-operation is not that good, as people don't really feel or see the benefits of Erin. Some are still of the view that it only benefits some of the community and not everyone, which is true, and we [are] not sure how to go around it. . . . With the Erin success we can currently only help some people during the hunting seasons, which is not enough . . . . [I]t is far from being sustainable. (L. Steenkamp, 15 November 2017)

Several times during my 2016 fieldwork, I heard criticisms of Erin Farm and the Bushman Raad as benefiting too few ǂKhomani at that time. Despite Luce's clear desire to share income and meat with, and create employment for, more of the community, there is only so much that smaller livelihood projects are currently able to do, particularly if, as in the case of Erin, there are still loans to be repaid. Still, according to Suzman (2017; citing Richard Lee's work among the Ju/'hoansi of Namibia), the expression of this kind of envy and selfishness from other community members has traditionally functioned in important ways among the San and other hunter-gatherers:

> In the case of small-scale hunter-gatherer societies, the sum of individual self-interests ultimately ensured the most equitable "distribution of the necessaries of life," and in doing so created the most sustainable economic model in modern *Homo sapiens* history. How this worked is best exemplified in the customary "insulting" of a hunter's meat. . . . Regardless of the size or condition of the carcass, those due a share of the meat would complain that the kill was trifling, that it was barely worth the effort of carrying it back to camp, or that there wouldn't be enough meat to go around. . . . Everyone knew the difference between a scrawny kill and a good one, of course, but nonetheless continued to pass insults even while they were busy filling their bellies with meat—the most highly prized of all foods.
>
> This behaviour was not limited to hunting. Similar insults were meted out to anyone who assumed airs and graces, encountered a windfall or got too big for their leather sandals. . . . They took careful note of what others ate, owned, received as gifts, and whether or not they were sufficiently generous in return. The net result was that everyone went to considerable lengths to avoid being singled out for selfishness or self-importance—so much so, indeed, that good hunters usually hunted less often than poor ones, even if they enjoyed it. Unsurprisingly, this created an atmosphere that was generally harmonious, co-operative, and in which even those with the natural charisma and character to lead did so only with great circumspection.

While, as Suzman argues, this kind of "insulting" culture can ultimately produce the cooperation and sustainability that Luce fears may be at risk among the ǂKhomani, it is also clear that this only works in "small-scale" societies.

In the situation that the ǂKhomani are now facing—a return to centralized, hierarchical leadership, decision-making, and financial-management structures without administrator support—the current, self-acknowledged inability of the Erin Farm managers to produce enough income and "the most highly prized of all foods" to share in "the most equitable" way across the community might put Erin at risk. In terms of future ǂKhomani community projects and land management more broadly, Luce Steenkamp reflected on the transition from a dedicated administrator back to a CPA Committee: "Government seems not to have a clear answer on the way forward and did not visit us [after] Terance's term came to an end" (June 16, 2021). Perhaps the illegal Miershoopan hunt in late 2020 will come to be understood as a violent expression of this uncertainty and the jealousy noted by Luce above. Those who view Erin as not "sufficiently generous" or as an inequitable, unwise, unsustainable use of "community" funds might also begin to use the new CPA Committee—who are once again being forced, by a combination of legal and community expectations, to attempt to reach community-wide consensus—to try to divert funding, deny approvals, and thus block the actions of the Traditional Land Ward Committee's skilled, "San-style" leaders.

## A POSSIBLE SOLUTION

While David Grossman characterized the 2014 appointment of Terance Fife as a fully empowered, properly funded ǂKhomani "administrator" as answered prayer, he also listed the "possibility of another corrupt CPA Committee being elected" as a major threat to Erin Farm (February 27, 2017). Because Terance always intended—as the CPA Act now mandates—for a new ǂKhomani CPA election to take place (November 16, 2018), it is also evident that the current ǂKhomani Constitution must be re-amended. I argue that this will be the surest way to provide space for small-group decision-making structures like the former Bushman Raad—which grew up quietly during the absence of a CPA Committee as an expression of the preferred "San-style" ǂKhomani leadership—to become more permanent facets of ǂKhomani governance moving forward. Although, as David pointed out, the CPA Act "makes provision for a CPA Committee," he further argued as follows:

> The . . . CPA Committee . . . should be an . . . oversight body, with responsibility to maintain assets and ensure they are used equitably. Decisions needed at Ward level should be devolved to Ward committees and made by the Ward committees. (Grossman, 1 March 2017)

Similarly, when asked about the potential impacts of a return to the old community-wide CPA Committee format post-administration, Luce Steenkamp made the following two statements:

> I don't think that would work. Everything would be fine if we could focus! It's worked now for years! Why can't we just let it keep going without making things more complicated? (27 August 2016)
>
> The Committee, I personally feel, are not ready to manage the community assets on their own, or . . . to make good decisions for the community when Terance's contract ends . . .. I am not sure if we are ready to get out of administration. (7 December 2020)

With these comments from David and Luce in mind, I recommended to Terance Fife, as part of the ǂKhomani conversations about creating a newly amended ǂKhomani Constitution, that they consider making the following (or similar) change to their current Constitution:

> [CURRENT]
>
> Clause 6.3.2. The functions of any such Ward Committee with regard to such Ward shall be:
>
> ...
>
> (e) to compile or have compiled, management plans for the various land uses and/or portions of land including budgets for the management, control and development of land and to submit same to the Management Committee for approval.
>
> [TO BE ADDED]
>
> In the spirit of ǂKhomani San traditions and the principle of equity, when such plans relate to smaller, Ward- or farm-based projects designed to generate or procure food, water, medicines, fuel, power, housing, infrastructure, income, other external donations, transportation, cultural enrichment, educational opportunities, or other benefits to one or more Ward farms, *such approval shall be given automatically* by the Management Committee to the individual Ward Committee, Clause 6.6 "subcommittee," or small-group organizers of such projects, unless or until the Management Committee, by [majority or two-thirds?] vote after all affected project participants have been given the opportunity to speak publicly, finds clear evidence of actual or imminent fraud, mismanagement, conflicts of interest, discrimination, abuse of authority, cruelty, illegal activities, or violations of Clause 17[28] or any other provisions of this Constitution.

According to Terance, the Constitutional-amendment process was to be introduced and outlined during the ǂKhomani Annual General Meeting on September 14, 2019, and as this process unfolded, my amendment idea—as well as those from other contributors, including the Legal Resources Center and the law firm Webber Wentzel[29]—were to be presented to, and discussed by, the ǂKhomani community (Fife, August 29, 2019). In March 2021, the ǂKhomani constitutional changes were "still at a consultative phase," and my proposed amendment was being reviewed by David Mayson at Phuhlisani, who was managing the constitutional change process at that time (Fife, March 24, 2021). As of August 2024, however, the ǂKhomani Constitution has not yet been amended, and meetings have been scheduled for ǂKhomani San living on the farms and in nine other settlements across the Northern Cape to continue—and attempt to finalize—these discussions (Pienaar, August 7, 2024). While clearly necessary for a decision as important as amending their

Constitution, the length and complexity of this ongoing process offer further evidence of the difficulty of reaching consensus among all ǂKhomani San.

In essence, the spirit of my proposed amendment would be to reduce the currently excessive requirements for community-wide agreement and CPA Committee input, authority, or approvals when smaller, "San-style" livelihood projects are involved. It would help to codify and enshrine the successful, behind-the-scenes leadership and decision-making processes that began to (re) emerge among the ǂKhomani while Peter Mokomele served as their "unofficial" administrator and could thus help to protect—and to encourage the launching of more—small-group ǂKhomani projects like Erin Farm. Indeed, Terance Fife's reactions to this proposed amendment were very positive. He noted that, "If you're not allowed to do something, you won't do it," and that this amendment would clearly help to "allow that [Bushman-Raad-style] initiative" to continue to spread among the ǂKhomani—"I like that!" (Fife, August 29, 2019).

With Erin Farm as the primary success story, the ǂKhomani have truly begun to flourish since 2008, when they were temporarily freed from the externally imposed CPA Committee structures. Unofficial, behind-the-scenes bodies, fora, and methods that quietly grew up during this time blossomed

**Figure 5.21 Zebra on an Area of Erin Farm that was Formerly Over-grazed by Sheep. Only Successful ǂKhomani Leadership, Negotiations, and Cooperation could have Made this Transformation Possible.** *Source*: Photo by the author.

and produced the first major successes on the ǂKhomani farms since 1999. These included both the Bushman Raad's Erin Farm operations and the ǂKhomani Stock-Farmers' Association's thorough, painstaking allocation of sheep "camps" for the ǂKhomani farmers. (See figure 5.21.)

## CONCLUSION

In an era when "Africa has the highest levels of severe food insecurity, affecting 27.4% of the population—almost four times that of any other region" (McVeigh 2017), "hundreds of millions of Africans are still without power or safe drinking water" (*The Economist*, November 11, 2017), and "nearly half (47%) of [South Africans] said they could not afford enough food" during the pandemic (*The Economist*, July 18, 2020), the ǂKhomani San today are striving to overcome these severe challenges with immense effort and courage. (See figure 5.22.)

As an internally conceptualized and driven venture, Dr. Grossman reported, "most [ǂKhomani] people who I have spoken to appear proud of their game ranch . . . . Prospects for the future? I remain very optimistic" (February 27, 2017).

**Figure 5.22 ǂKhomani Goats Grazing near Power Lines, which Provide Electricity to the Local School, Police Station, Hotel, and Shop, but which Bypass all ǂKhomani Homes and budding Livelihood Ventures. The ǂKhomani have Brought in Generators and Solar Panels where Possible, but These will Not be Sufficient in the Long Run**. *Source*: Photo by the author.

Indeed—despite the unclear motivations of the ǂKhomani Traditional Leader during his 2020 "invasion" of Miershoopan Farm and his community-blocked attempt at a further illegal hunt on Erin Farm itself—the general cohesion seen among these formerly dispersed San is higher than it has been since the arrival of South African democracy in 1994 and perhaps since the beginning of the Park evictions in 1931. In fact, the lack of forced, fully functioning "community" structures after 2008 led to more identification among ǂKhomani farm residents with the idea of a true "ǂKhomani *community*" than any of the CPA Act's imposed requirements—based on legislated ideals of representative democracy, communal ownership, or shared rules—had been able to engender during the first decade of ǂKhomani resettlement in the Kalahari (see also Puckett 2018). Not only are members of both "traditional" and stock-farming ǂKhomani families working together on Erin Farm, but this new "community" spirit (which was merely imaginary during the 1990s land-claim process) is manifesting itself in a variety of ways, including renewed interest in the endangered traditional ǂKhomani language, N/uu. In 2023, for example, the Kalahari Peoples Fund received several requests from ǂKhomani community members for funding to pay a N/uu language instructor to teach young ǂKhomani children in their Andriesvale "creche," or daycare center.

So, clearly, with the level of support, training, funding, current success, and community pride involved, a lot is at stake when it comes to Erin Farm. Maintaining the Indigenous small-group structures that have grown up behind the scenes during "administration" will be critical, as a return to ideals, systems, and structures imposed from the outside will threaten the activity levels and real progress seen on the ǂKhomani farms since 2008. Empowering and incorporating these small-group structures into the required CPA Act and ǂKhomani Constitution governance systems will allow the ǂKhomani to build on the inspiring results from Erin Farm and should provide a platform for new San-style experts, advisers, and leaders to emerge. (See figure 5.23.)

Dr. Grossman wrote that a key ingredient in the initial success of Erin Farm was "determination by the people to show that THEY CAN" (February 27, 2017). And despite the ongoing risks, criticisms, and uncertainties, Erin Farm should provide—and has begun to provide—an inspiring success story, allowing other typically "circumspect" ǂKhomani leaders to begin to emerge, share their ideas, talk about the future with optimism, and perhaps begin to fill the significant void left by the deaths of Dawid Kruiper and Andries Steenkamp.

I argue that the willingness to stand up to their current Traditional Leader during the illegal hunting incident, as the ǂKhomani, in Terance's words, "remained resolute and intervened . . . eventually stopping [the illegal hunting] activities" before the group reached Erin Farm (Fife, October 12, 2020), is further, encouraging evidence that—despite livelihood, leadership, and

**Figure 5.23 Erin Farm is Proving to be a Priceless Asset for the ǂKhomani San, not only as a Source of Income, but also as a Space to Reconnect with their own History, Skills, Stories, Knowledge, and Identities. After Centuries of Persecution across Southern Africa, Success Stories like Erin may help Younger San begin to Step Forward with Confidence, to Speak Out, Share Ideas, Pursue their own Goals, Contribute to their Communities, and Inspire Others.** *Source*: Photo by the author.

resource disputes that may arise in the community—the ǂKhomani San people clearly recognize (and are willing to defend!) the great asset that they have created on Erin Farm.

To conclude, it appears that a tourism venture (designed primarily for non-community members) might ultimately provide the greatest benefits to the community itself—income, food, employment, skills transfer, pride, activity, and community cohesion—of any program or project attempted so far on the lands of the ǂKhomani San. With the success of Erin, the ǂKhomani—despite being labeled by the government as previously "failing" to govern themselves—are finally starting to see the cultural "rebirth" promised by the government via ǂKhomani land restitution back in 1999. And during the period from 2008 to 2021 at least, this rebirth was able to take place on the ǂKhomani San's own terms. The impacts of the opportunity under "administration" for the ǂKhomani, as Peter noted above, to "define themselves" and "find their own arrangements" seem to illustrate perfectly the statement made by James Suzman (2017) that, today, the San people "are engaging with modernity with the hands and minds of hunter-gatherers."

In discussing what it means to be a hunter-gatherer, former Bushman Raad leader "Oupa Jan" Pietersen explained, "If we speak about tradition, it is all about nature. Nature can teach you. It can also heal you" (September 1, 2009). Perhaps, in a sense, the ability of the ǂKhomani—particularly after 2008—to return to what is natural to them, to express their traditions, to spend time on their new lands in their own ways (stock-farming, craft-making, medicine-gathering, wood-collecting, guiding, tracking, hunting, or just walking, observing, or being) has given them the time, as Peter suggested, to find—or remember—their own "rules of engagement," to sit and talk and heal and then to plan and then to act. But it started with the opportunity to meet and talk and decide in their own ways.

## NOTES

1. "Wildsplaas" is the Afrikaans word for *game farm*.
2. An amended Restitution Act was passed in 2014. However, the primary purpose of this new law was to re-open the claims process and extend the land-claim filing deadline to June 2019 (see Restitution of Land Rights Amendment Act, No. 15 of 2014).
3. South Africa's DLA was later renamed the Department of Rural Development and Land Reform and then more recently the Department of Agriculture, Land Reform, and Rural Development (DALRRD). However, the ǂKhomani San continue to use the names "Department of Land Affairs" and "DLA" today.
4. Low-cost houses built by the government under South Africa's Reconstruction and Development Program, typically for households (often extended families) living off social and disability grants and earning 3,500 Rand or less per month (Greyling 2009), which equates to just over US$ 6/day in November 2023.
5. "Oupa" meaning *grandfather* and "Oom" meaning *uncle* are both Afrikaans terms used by the ǂKhomani San, in this case, for respected experts or elders, regardless of actual age or familial relationship.
6. Shaka kaSenzangakhona (c. 1787 to 1828), the great unifying king of the Zulu nation.
7. This settlement created, empowered, and fully funded an official "administrator" role (Terance Fife of Cape Town) for the ǂKhomani San for the first time. From 2008 to 2013, Peter Mokomele filled this role, but very often conducted this work on his own time, with his own vehicle and personal funds.
8. The term "Raad" is Afrikaans for *council*. This "council," however, lacked the size, structure, formality, and hierarchies that would typically be encountered in such a body in the West.
9. As of 2012, when the pilot hunting season was launched, there were no remaining residents on the main body of Erin Farm.
10. From Tswalu Kalahari Game Reserve, run by the Oppenheimer family, to the east of the ǂKhomani farms, near the town of Van Zylsrus. See https://www.tswalu.com/.
11. See Puckett (2018) for a discussion of "traditional" vs. "Western" ǂKhomani. I argue that this divide relates far more to outward livelihood choices and focus areas

than to any differences in inward-facing hunter-gatherer identities, which appear to be fairly consistently "traditional" across the ǂKhomani, regardless of location or occupation. *Cf.* Robins 2001b:841–42.

12. GhoenasKraal Opleidingsentrum website: http://www.ghoenaskraal.co.za/wmenu.php (Afrikaans only).

13. Example of certificate: South African Qualifications Authority's certificate in "Animal Production." http://regqs.saqa.org.za/viewQualification.php?id=48970. The South African Qualifications Authority administers the National Qualifications Framework, or "NQF," under South Africa's National Qualifications Framework Act 67 of 2008.

14. According to the land-claim attorney for the ǂKhomani San, Dr. Roger Chennells (December 13, 2009), although Miershoopan had been an original land-claim farm designated for "traditional" use by the ǂKhomani, a drafting error by the Legal Resources Center during the amendment of the ǂKhomani CPA Constitution (adopted on May 4, 2002) mistakenly re-assigned Miershoopan to those ǂKhomani seeking primarily stock-farming livelihoods, and it remains so allocated in their current Constitution today. However, as of 2024, conversations are taking place among the ǂKhomani San about moving ahead with Miershoopan Farm as a second "wildsplaas."

15. This is part of Louis Liebenberg's important South African tracker-certification program at CyberTracker Conservation NPC. See https://www.cybertracker.org/tracking/evaluations.

16. Among authors, the traditional structure of San small-group organization is typically termed the "band" or "camp" and is often tied to a particular piece of land and one or more water sources (see, e.g., Lee 1998b:77; Marshall 1998:350; Hitchcock & Biesele 2000:15–16; and Smith et al. 2000:70, 73). Among the ǂKhomani, I did not hear any use of the term "band," but the term "camp" is indeed used by the ǂKhomani in a geographical context, designating the small plots of grazing land that are allocated to individual families.

17. "Biltong" is roughly equivalent to our *beef jerky*, but in South Africa is made from what they call "venison" there, usually kudu or springbok meat.

18. Following the 1998/1999 discussion period and the 2008/2009 renewed planning period, Erin Farm was indeed physically developed over a three-year period: 2010, 2011, and 2012, when Erin's pilot hunting season opened.

19. For discussions of San adaptability over time, including taking advantage of changing circumstances, landscapes, and local markets, see Hitchcock & Biesele 2000:6–7; Smith et al. 2000:82; Taylor 2002:471, 475, 481; Tomaselli 2002:204; and Barnard 2007:6.

20. The nonprofit African Safari Lodge Foundation (or simply "African Safari Foundation") is based in Johannesburg and works with rural communities across southern Africa in a variety of capacities, including providing tourism planning and nature-guide training in cooperation with the South African Department of Tourism. This work is funded in part by grants from the Ford Foundation and the South African Lottery and falls under the framework developed by the South African government's "Quality Council for Training and Occupations." See http://www.asl-foundation.org/.

21. See www.khomanisan.com/erin-game-ranch/.

22. See http://www.jammainternational.com/f1x6c8ervt283gorpyis4m4kqpzo54, including information on JAMMA funding for a new ǂKhomani school.

23. See http://lashlerouxsafaris.co.za/.

24. See ǂKhomani Symbolic-Rights Area ("S-Zone") and Contract Park (Heritage Park and "V-Zone") in the Park on map, Figure 5.1.

25. ǂKhomani elders were once able to pass down their "knowledge of nature and *veld* and the historical and symbolic places" to the next generation through the *veldskool* (field school) founded by Dawid Kruiper for the ǂKhomani children at two locations inside the Park (Bobbejaanskop and Tiermelk) in May 2009 (Pietersen, September 1, 2009 and January 16, 2010). Sadly, the school has since closed, as its funding—at various times from the Peace Parks Foundation, BMZ (the German Federal Ministry of Economic Cooperation and Development), and Rotary International—all came to an end in 2018 (L. Steenkamp, January 11, 2021).

26. See "HEADLINES" section of http://www.khomanisan.com/.

27. Section 6.3 of the current ǂKhomani Constitution defines the Traditional Land Ward Committee as consisting of members elected from the populations of Erin and Witdraai Farms, as automatically including the Traditional Leader of the ǂKhomani San, and as being tasked, in part, with the maintenance of ǂKhomani San traditions.

28. Clause 17 of the current ǂKhomani Constitution is entitled, "Code of conduct for members of the Management Committee, subcommittees, and other appointments" and lays out guidelines, rules of behavior, and accountability requirements for those chosen to represent, and to help meet the needs of, the ǂKhomani community.

29. Webber Wentzel's official website: https://www.webberwentzel.com/Pages/default.aspx.

## REFERENCES

African Safari Foundation. http://www.asl-foundation.org/. Accessed 13 January 2018.

Barnard, Alan. *Anthropology and the Bushman*. Oxford: Berg, 2007.

Barnard, Alan. "The Foraging Mode of Thought." In *Self- and Other-Images of Hunter Gatherers*, edited by Henry Stewart, Alan Barnard, and Keiichi Omura, 5–24. Senri Ethnological Studies No. 60. Osaka: National Museum of Ethnology, 2002.

Blurton Jones, Nicholas. "Why Do So Few Hadza Farm?" In *Why Forage? Hunters and Gatherers in the Twenty-First Century*, edited by Brian F. Codding and Karen L. Kramer, 113–136. Santa Fe: School for Advanced Research Press, 2016.

Bradstock, Alastair. "Land Reform and Livelihoods in South Africa's Northern Cape Province." *Land Use Policy* 23 (2006): 247–259.

Bradstock, Alastair. "Changing Livelihoods and Land Reform: Evidence from the Northern Cape Province of South Africa." *World Development* 33, no. 11 (2005): 1979–1992.

Buntman, Barbara. "Travels to Otherness: Whose Identity Do We Want to See?" In *Self- and Other-Images of Hunter Gatherers*, edited by Henry Stewart, Alan

Barnard, and Keiichi Omura, 65–84. Senri Ethnological Studies No. 60. Osaka: National Museum of Ethnology, 2002.
Chennells, Roger. Personal communication, 26 October 2020.
Chennells, Roger. Interview with the author. Kimberley, South Africa: 13 December 2009. [Recording in possession of author.]
Chennells, Roger. *Report on the Land Rights of the ǂKhomani San Community*. Office of the Commission on Restitution of Land Rights, 2006.
Chennells, Roger. "The Khomani San Land Claim." *Cultural Survival Quarterly* 26, no. 2 (2002): 51.
"Climate Change in Africa: Scrambled Skies." *The Economist* 448, no. 9359 (2023): 40.
Commission on Restitution of Land Rights. *Restitution of Land Rights Amendment Act, No. 15 of 2014*. Cape Town: Government Gazette 589, no. 37791 (2014).
Commission on Restitution of Land Rights. *Restitution of Land Rights Act, No. 22 of 1994 as Amended*. Pretoria: Department of Land Affairs, 2004.
*Communal Property Associations Act, No. 28 of 1996*. President's Office, No. 849 (22 May 1996). http://www.caledonia.org.uk/land/cpaact.htm.
Cripps, Karla. *CNN*: "UNESCO 2017: Which sites have been added to the World Heritage List?" 10 July 2017. http://www.cnn.com/travel/article/unesco-2017-world-heritage-sites/index.html. Accessed 7 January 2018.
de Soto, Hernando. *The Mystery of Capital: Why Capitalism Triumphs in the West and Fails Everywhere Else*. New York: Basic Books, 2000.
Fife, Terance. Personal communication, 25 March 2021.
Fife, Terance. Personal communication, 24 March 2021.
Fife, Terance. Personal communication, 12 October 2020.
Fife, Terance. Interview with the author. Andriesvale, South Africa: 29 August 2019. [Hand-written notes in possession of author.]
Fife, Terance. Interview with the author. Cape Town, South Africa: 16 November 2018. [Hand-written notes in possession of author.]
GhoenasKraal Opleidingsentrum. http://www.ghoenaskraal.co.za/wmenu.php, (Afrikaans only). Accessed 16 January 2018.
Giraudo, Rachel F. "Performing Our Past to Secure Our Future: A Look at San-Owned Cultural Tourism in the Kalahari." In *Research and Activism among the Kalahari San Today: Ideals, Challenges, and Debates*, edited by R. Fleming Puckett and Kazunobu Ikeya, 215–30. Senri Ethnological Studies 99. Osaka: National Museum of Ethnology, 2018.
Grant, Julie. "Rural Development in Practice: The Experience of the ≠Khomani Bushmen in the Northern Cape, South Africa." PhD diss., University of Edinburgh, 2011.
Greyling, Carin. "The RDP Housing System in South Africa." B.Sc. (Hons) thesis, University of Pretoria, 2009.
Grossman, David. Personal communication, 1 March 2017.
Grossman, David. Personal communication, 27 February 2017.
Grossman, David. Interview with the author. Andriesvale, South Africa: 20 January 2010. [Typed notes in possession of author.]

Guenther, Mathias G. Discussion: "The Concept of Indigeneity." *Social Anthropology* 14, no. 1 (2006): 17–19.

Guenther, Mathias G. "Ethno-tourism and the Bushmen." In *Self- and Other-Images of Hunter Gatherers*, edited by Henry Stewart, Alan Barnard, and Keiichi Omura, 47–64. Senri Ethnological Studies No. 60. Osaka: National Museum of Ethnology, 2002.

Hitchcock, Robert K. and Megan Biesele. "Introduction." In *Hunters and Gatherers in the Modern World: Conflict, Resistance, and Self-Determination*, edited by Peter P. Schweitzer, Megan Biesele, and Robert K. Hitchcock, 1–27. New York: Berghahn Books, 2000.

Hitchcock, Robert K. and Maria Sapignoli. "Twenty-First-Century Hunting and Gathering among Western and Central Kalahari San." In *Why Forage? Hunters and Gatherers in the Twenty-First Century*, edited by Brian F. Codding and Karen L. Kramer, 89–111. Santa Fe: School for Advanced Research Press, 2016.

JAMMA International. www.jammainternational.com/f1x6c8ervt283gorpyis4m4kqpzo54. Accessed 16 January 2018.

ǂKhomani San. "Erin Game Ranch." www.khomanisan.com/erin-game-ranch/. Accessed 6 January 2018.

ǂKhomani San. "Friends of the Khomani San." www.khomanisan.com/foks/. Accessed 15 January 2018.

ǂKhomani San. "HEADLINES." www.khomanisan.com/. Accessed 31 July 2019.

Koot, Stasja. "The Bushman Brand in Southern African Tourism: An Indigenous Modernity in a Neoliberal Political Economy." In *Research and Activism among the Kalahari San Today: Ideals, Challenges, and Debates*, edited by R. Fleming Puckett and Kazunobu Ikeya, 231–250. Senri Ethnological Studies 99. Osaka: National Museum of Ethnology, 2018.

Kuper, Adam. "The Return of the Native." *Current Anthropology* 44, no. 3 (2003): 389–402.

Lash le Roux Safaris. http://lashlerouxsafaris.co.za/. Accessed 6 January 2021.

"The leapfrog model." In "Special Report: Technology in Africa." *The Economist* 425, no. 9066 (2017): 3–5.

Lee, Richard B. "'In the bush the food is free': The Ju/'hoansi of Tsumkwe in the Twenty-First Century." In *Why Forage? Hunters and Gatherers in the Twenty-First Century*, edited by Brian F. Codding and Karen L. Kramer, 61–87. Santa Fe: School for Advanced Research Press, 2016.

Lee, Richard B. (1998a). "Introduction." In *Kalahari Hunter-Gatherers: Studies of the !Kung San and Their Neighbors*, edited by Richard B. Lee and Irven DeVore, 1–24. Cambridge: Harvard University Press, 1998.

Lee, Richard B. (1998b). "!Kung Spatial Organization: An Ecological and Historical Perspective." In *Kalahari Hunter-Gatherers: Studies of the !Kung San and Their Neighbors*, edited by Richard B. Lee and Irven DeVore, 73–97. Cambridge: Harvard University Press, 1998.

Liebenberg, Louis. Tracker-certification program at CyberTracker Conservation NPC. https://www.cybertracker.org/tracking/evaluations. Accessed 7 January 2018.

Marshall, Lorna. "Sharing, Talking, and Giving: Relief of Social Tensions among the !Kung." In *Kalahari Hunter-Gatherers: Studies of the !Kung San and Their Neighbors*, edited by Richard B. Lee and Irven DeVore, 349–72. Cambridge: Harvard University Press, 1998.

Maslow, A. H. "A Theory of Human Motivation." *Psychological Review* 50, no. 4 (1943): 370–96.

McVeigh, Karen. "'Alarm bells we cannot ignore': world hunger rising for first time this century." *The Guardian*, 15 September 2017. https://www.theguardian.com/global-development/2017/sep/15/alarm-bells-we-cannot-ignore-world-hunger-rising-for-first-time-this-century. Accessed 14 January 2018.

"Measuring the poverty pandemic." *The Economist* 436, no. 9203 (2020): 38–39.

Mokomele, Peter. Interview with the author. Kimberley, South Africa: 2 September 2016. [Recording in possession of author.]

Mokomele, Peter. Interview with the author. Kimberley, South Africa: 15 September 2009. [Recording in possession of author.]

Oliver, Roland. "Bantu Genesis: An Inquiry into Some Problems of Early Bantu History." *African Affairs* 65, no. 260 (1966): 245–258.

O'Malley, Pat. "Indigenous Governance." *Economy and Society* 25, no. 3 (1996): 310–326.

Pienaar, Dirk. Personal communication, 7 August 2024.

Pienaar, Dirk. Personal communication, 3 November 2023.

Pienaar, Dirk. Personal communication, 20 October 2023.

Pienaar, Dirk. Interview with the author. Erin Farm, South Africa: 30 June 2023. [Hand-written notes in possession of author.]

Pienaar, Dirk. Interview with the author. Erin Farm, South Africa: 30 August 2019. [Hand-written notes in possession of author.]

Pienaar, Dirk. Personal communication, 4 December 2017.

Pietersen, Jan Johannes. Interview with the author. Sonderwater Farm, South Africa: 16 January 2010. [Recording in possession of author.]

Pietersen, Jan Johannes. Interview with the author. Witdraai Farm, South Africa: 1 September 2009. [Recording in possession of author.]

Puckett, R. Fleming. "'The Space to Be Themselves': Confronting the Mismatch Between South Africa's Land Reform Laws and Traditional San Social Organization among the ǂKhomani." In *Research and Activism among the Kalahari San Today: Ideals, Challenges, and Debates*, edited by R. Fleming Puckett and Kazunobu Ikeya, 283–342. Senri Ethnological Studies 99. Osaka: National Museum of Ethnology, 2018.

Puckett, Robert L. Personal communication, 11 January 2018.

Robins, Steven. (2001a). "Part I: South Africa." In *An Assessment of the Status of the San in South Africa, Angola, Zambia and Zimbabwe*, edited by Steven Robins, Elias Madzudzo, and Matthias Brenzinger, ix–51. Series Report no. 2 of 5 in *Regional Assessment of the Status of the San in Southern Africa.* Windhoek: Legal Assistance Centre, 2001.

Robins, Steven. (2001b). "NGOs, 'Bushmen' and Double Vision: The khomani San Land Claim and the Cultural Politics of 'Community' and 'Development' in the Kalahari." *Journal of Southern African Studies* 27, no. 4 (2001): 833–853.

Saugestad, Sidsel. "Contested Images: 'First Peoples' or ''Marginalized Minorities'' in Africa?" In *Africa s Indigenous Peoples: First Peoples or Marginalized Minorities ?*, edited by Alan Barnard and Justin Kenrick, 299–322. Edinburgh: University of Edinburgh Centre of African Studies, 2001.

*Second Constitution of the ǂKhomani-San Communal Property Association.* Uitkoms Farm, South Africa: ǂKhomani-San Association for Communal Property, 4 May 2002.

Smith, Andy, Candy Malherbe, Mat Guenther, and Penny Berens. *The Bushmen of Southern Africa: A Foraging Society in Transition.* Cape Town: David Philip Publishers, 2000.

South African Qualifications Authority. "National Certificate: Animal Production." http://regqs.saqa.org.za/viewQualification.php?id=48970. Accessed 7 January 2018.

Steenkamp, Andries. Interview with the author. Erin Farm, South Africa: 23 January 2010. [Recording in possession of author.]

Steenkamp, Andries. Interview with the author. Erin Farm, South Africa: 8 September 2009. [Recording in possession of author.]

Steenkamp, Luce. Personal communication, 28 January 2023.

Steenkamp, Luce. Personal communication, 21 February 2022.

Steenkamp, Luce. Personal communication, 27 January 2022.

Steenkamp, Luce. Personal communication, 16 June 2021.

Steenkamp, Luce. Personal communication, 17 March 2021.

Steenkamp, Luce. Personal communication, 11 January 2021.

Steenkamp, Luce. Personal communication, 7 December 2020.

Steenkamp, Luce. Personal communication, 16 October 2020.

Steenkamp, Luce. Personal communication, 15 November 2017.

Steenkamp, Luce. Personal communication, 1 March 2017.

Suzman, James. "Why 'Bushman banter' was crucial to hunter-gatherers' evolutionary success." *The Guardian*, 29 October 2017. https://www.theguardian.com/inequality/2017/oct/29/why-bushman-banter-was-crucial-to-hunter-gatherers-evolutionary-success?CMP=share_btn_link. Accessed 14 January 2018.

Sylvain, Renée. "Disorderly Development: Globalization and the Idea of 'Culture' in the Kalahari." *American Ethnologist* 32, no. 3 (2005): 354–370.

Sylvain, Renée. "'Land, Water, and Truth': San Identity and Global Indigenism." *American Anthropologist* 104, no. 4 (2002): 1074–1085.

Taylor, Michael. "The Shaping of San Livelihood Strategies: Government Policy and Popular Values." *Development and Change* 33, no. 3 (2002): 467–488.

Tomaselli, Keyan. "'. . .We have to Work with Our Own Heads' (/Angn!ao): San Bushmen and the Media." *Visual Anthropology* 15, no. 2 (2002): 203–220.

Tswalu Kalahari Game Reserve. https://www.tswalu.com/. Accessed 6 January 2018.

Useb, Joram. "'One Chief Is Enough!': Understanding San Traditional Authorities in the Namibian Context." In *Africa's Indigenous Peoples: 'First Peoples' or 'Marginalized Minorities'?*, edited by Alan Barnard and Justin Kenrick, 15–29. Edinburgh: University of Edinburgh Centre of African Studies, 2001.

Van der Westhuizen, Vinkie. Interview with the author. Andriesvale, South Africa: 8 September 2009. [Recording in possession of author.]

Von Bremen, Volker. “Dynamics of Adaptation to Market Economy among the Ayoréode of Northwest Paraguay.” In *Hunters and Gatherers in the Modern World: Conflict, Resistance, and Self-Determination*, edited by Peter P. Schweitzer, Megan Biesele, and Robert K. Hitchcock, 275–286. New York: Berghahn Books, 2000.

Webber Wentzel. https://www.webberwentzel.com/Pages/default.aspx. Accessed 22 September 2019.

Wenzel, George W. “Inuit Culture: To Have and Have Not; or, Has Subsistence Become an Anachronism?” In *Why Forage? Hunters and Gatherers in the Twenty-First Century*, edited by Brian F. Codding and Karen L. Kramer, 43–60. Santa Fe: School for Advanced Research Press, 2016.

Woodburn, James. “Egalitarian Societies.” *Man* (new series) 17, no. 3 (1982): 431–451.

## APPENDIX 1.

According to Andries Steenkamp (personal communication, January 23, 2010), the 1,024 ǂKhomani San living in the region at that time were distributed as follows:

360 ǂKhomani living on or adjacent to the new ǂKhomani lands:
- Six land-claim farms – 218;
- Settlement of Welkom, outside the Park – 82;
- Town of Askham – 60.

664 ǂKhomani remaining in more distant Northern Cape towns:
- Rietfontein – 180;
- Loubos – 102;
- Mier I and II – 72;
- Philandersbron – 80;
- Noenieput – 30;
- Upington – 200.

These figures confirm Bradstock’s findings (2005 & 2006) that many ǂKhomani have, for a variety of reasons, been unable to relocate to the new ǂKhomani farms and live there.

## Chapter 6

# The Mayanization of the Ch'orti'

## *Indigenous Identity amid Tourism-Based Development Initiatives in Copán, Honduras*

Fredy Rodriguez-Mejia

> *I submitted a proposal [to the World Bank tourism project] . . . but there was nothing for me . . . because of the discrimination that exists against us indigenous people . . . only three Ch'orti' communities benefited [from these projects] . . . the opportunities were given to those who already had money . . . giving more money to those who own the souvenir shops and the hotels . . . they benefited from those funds that were meant for indigenous people.*
>
> Doña Carmen Mendoza (pseudonym),
> Ch'orti' Maya leader and activist.

The 1990s was a crucial decade for Indigenous groups in Honduras in matters of political mobilization, ethnic recognition by the state, tourism development projects, and cultural revitalization initiatives. With the election of President Rafael Leonardo Callejas (1989), Honduras formally entered the "neoliberal march," giving rise to several socioeconomic reforms and unanticipated political mobilizations. Prior to securing the Honduran presidency, Callejas had already negotiated multiple economic restructurings in partnership with the World Bank, USAID, and the Inter-American Development Bank. Through his "Structural Adjustment" programs, Callejas devalued the Honduran currency (Lempira) by 50 percent, augmented consumption taxes, eliminated the control of prices, reduced trade tariffs, reduced the workforce for the public sector, and supported privatization (Brondo 2013, 40; Hoksbergen & Espinoza Madrid 1997, 41). Other crucial reforms spearheaded by Callejas included strong support for transnational investors, many of whom focused on "modernizing" the agricultural sector and reinforcing the tourism economy in regions such as the Caribbean coast and rural archaeological regions such as Copán (Brondo 2013, 41; Barahona 2005).

It is within this neoliberal context that sustainable development surged as an attractive approach to move Honduras in the direction of democratic stability and economic recovery following more than a century of military and dictatorship regimes. With a population of 9.4 million as of 2019, Honduras remained the second poorest country in Central America, with close to 65 percent of its population living in poverty. Even though from 2010 to 2017 Honduras experienced an economic growth of 3.1 percent to 4.0 percent, the country still faces precarious economic conditions fueled by government corruption, political uncertainty, and heightened violence (Central Intelligence Agency 2020). Still, data from the U.N. Human Development Index (HDI) from 1990 to 2018 positions Honduras in a "medium human development" category (ranking it 132 out of 189 countries). The HDI measures countries' long-term growth in areas such as: life expectancy, years of schooling, and standard of living. For the period 1990–2018, Honduras' HDI value increased by 22.6 percent, showing the following values: (1) increase in life expectancy by 8.4 years, (2) increase in years of schooling by 1.6 years, and (3) increase in Gross National Income by 40.7 percent (United Nations Human Development Report 2019, 2–3). Amid this moderate development progress, however, the rural sector, especially Indigenous populations, continues to be affected disproportionately in economic issues. Thus, the tourism industry surged as a national approach supported by transnational funding agencies (e.g., the World Bank) with an eye toward poverty reduction (Ferguson 2010).

The case of the Ch'orti' Maya of Copán (in western Honduras) reveals an important dimension of the relationship that emerged between the state, Indigenous people, and non-Indigenous society following different development strategies. These strategies include "tourism microenterprise funding" (Ferguson 2010) and identity-based livelihood strategies such as heritage tourism (Mortensen 2005). After the political mobilization of the Ch'orti' Maya gained national and international attention in the late 1990s, Indigenous people were gradually integrated into tourism-based development initiatives by both state ministries and non-Indigenous businesspeople. In turn, tourism-related projects created different opportunities for people to mobilize their identities in creative ways. In this chapter, using different ethnographic vignettes, I explore the emergence of heritage and identity-based practices for tourism consumption and livelihood strategies, and how these practices relate to Indigenous activism. I emphasize how Indigenous people creatively use their identities and political activism to compete with non-Indigenous society as identity becomes commoditized. I will argue that while Indigenous identity becomes a marketable product used in multiple avenues, it is mostly non-Indigenous people who possess the economic capital to draw financial benefits. At the same time, some of these practices (e.g., theater performances) are starting to play a role in community-based traditions (e.g., agrarian festivals) as a way to reinvigorate Indigenous identity in places not accessible to non-Indigenous entrepreneurs.

## BACKGROUND AND METHODS

The municipal region of Copán Ruinas encompasses a small town of about 42,000 people with close to 100 different Indigenous villages located on the western side of Honduras, twelve kilometers from the Guatemalan border. Copán Ruinas is one of twenty-three municipal regions located within the department (or province) of Copán. The capital of the department is called Santa Rosa de Copán, and people often confuse this city with Copán Ruinas. Many people also refer to Copán Ruinas simply as Copán. I will do the same for the purposes of this chapter. Traditionally, people from Copán have been referred to as *copaneco/a* and people from the villages as *indigena*. Most *copaneco/a* people consider themselves *mestizo* (of mixed ethnicity and race), though a few families are of direct European descent. Labels among Copán residents are rarely used; however, most people from the town do not consider themselves Indigenous and draw identity divisions based on the geographic separation that exists between Copán and the villages. Geographic boundaries, however, are not enough to create an Indigenous/non-Indigenous dichotomy as some Indigenous families now live in some parts of Copán, and some non-Indigenous families have chosen to live in some villages traditionally inhabited by Indigenous people. Although I recognize the importance of ethno-racial subjectivities, for the sake of clarity in some of the arguments in this chapter, I will use the term non-Indigenous to refer to *mestizo* residents and people of direct Spanish descent.

Because of the growth of tourism and the archaeology industry (Mortensen 2005), Copán has become an attractive destination for travelers. Thus, being a native or non-native of Copán (and its villages) has a lot of implications for people's opportunities to work in the tourism industry and establish land and identity claims. However, the history of Copán and its people has been a difficult one to trace. With the exception of some colonial documents, little historical material exists about a single place by the name of "Copán" until it was established as a municipality in 1893. Feldman's (2009) historical reconstruction of the area comprises perhaps the most comprehensive depiction of what and where Copán was since the arrival of the Spanish to Honduras (1524) and its inhabitants. Instead of writing about a single locality, Feldman describes "the many Copáns" that have existed throughout history. First, the society that ruled the region during what has been scholarly known as the Lowland Classic Maya period (A.D. 250–900) (Andrews & Fash 2005). Though the site had been abandoned by the time the Spanish arrived in Honduras, many of the people who lived under the rule of the Classic Maya Copán dynasty remained in the area.

During the time of the European colony, well-known colonial documents such as Diego Garcia de Palacio's letter to King Phillip II (1576) talk about Copán in relation to the Maya ruins. Garcia de Palacio writes of the region as being a place formerly controlled by the Yucatán (southern Mexico) where people speak a language intelligible to those who live in the Yucatán (Herranz Herranz 1994, 89). During the time of the colony, the region of Copán was occupied by conquistador Pedro de Alvarado. In 1530, De Alvarado defeated Copán's Indigenous leader by the name of Copán Calel (whom the town and department are named after) (Garcia de Palacio 1576 [1985, 49]). For hundreds of years, the region of Copán lived under the Spanish forced labor systems such as *La Encomienda.* Through the *Encomienda,* a Spanish conquistador was assigned a set of restricted rights to property and Indigenous labor granted by the Spanish monarchs. Indigenous people had to pay tribute to their corresponding *encomendero* in the form of work, money, metals, and other goods. In return, the Indigenous people would receive "protection" and be indoctrinated into the Catholic faith. *Encomenderos* were also responsible for protecting the parcels of land assigned to them and also paying taxes to the Crown (Yeager 1995:843).

By the time Honduras achieved its independence in 1821, huge parcels of land were privatized, leading to the creation of massive estates or *haciendas* in Copán owned by a handful of families. Following the defeat of Copán Calel, land was transferred to its assigned *encomendero*—Diego Pineda de la Peña, until the land became property of the Republic of Honduras in the 1800s. Immigration of *mestizo* and *criollo* (direct Spanish descent) families to the area contributed to the growth of Copán by the end of the nineteenth century and the beginning of the twentieth century. Key contributors to the growth of Copán were the archaeological expeditions starting in the 1890s sponsored by Harvard's Peabody Museum and the growth of the tobacco industry (Rodriguez-Mejia 2011). Six *criollo* families gained control of most lands in the region of Copán. With the help of the Catholic Church, some Indigenous villages retained access to their land, but the vast majority were absorbed as peons.

By the 1950s, landowners had bought the rest of the land extending into Guatemala. The Indigenous villages that were established on some of these lands were displaced and further absorbed as workers of the *haciendas* (Martinez Perdomo 1997; Chernier et al. 1999:224). In the

1980s, the Ch'orti' Maya began to participate in class-based *campesino* (peasant) activism with minimal success. However, in the 1990s, with the help of anthropologists, transnational activists, and the support of non-governmental organizations, Indigenous communities established the first Indigenous council and officially began mobilizing as the Ch'orti' Maya ethnic group to demand land and ethnic recognition from the Honduras state. Although since the 1990s, not all demands have been met by the state, the Ch'orti' Maya have advanced many of their activism goals, including communal titles, ethnic recognition, the establishment of two Indigenous councils, cultural revitalization initiatives (e.g., language and agrarian rites), and education support for Indigenous teachers. The findings discussed in this chapter draw from some of the activism outcomes of the Ch'orti' Maya in combination with the growth of the tourism industry in Copán.

I conducted different phases of ethnographic research in Copán between 2008 and 2013. The bulk of the data was collected between June 2012 and April 2013 in the town of Copán and three Indigenous villages: La Pintada, El Ranchito (pseudonym), and San Rafael. Less than 700 people live in each of these villages. It is important to point out that this chapter draws from interviews, participant observation, and informal conversations over the years with a portion of the individuals involved in my larger dissertation research project. For the bulk of my dissertation work, I interviewed eighty-nine people from the villages averaging forty-one years of age. Forty-seven of the participants were men, and forty-two of the participants were women. Of the eighty-nine people interviewed in the villages, 69 percent claimed to be "Ch'orti' Maya," 25 percent were not familiar with the concept of ethnicity, and 6 percent claimed to be "Indigenous." About 14 percent of the participants spoke some Ch'orti' Maya and Spanish, and the other 86 percent spoke Spanish only. I also interviewed twelve people from the town of Copán. Ten out of the twelve participants claimed to be *mestizos* (mixed ethnicity and race), and two claimed to be both *mestizo* and Indigenous.

My research instruments included paper and pencil demographic surveys, individual interviews, focus groups, and participant observation. All interviews were conducted in Spanish and audio recorded with the permission of the participants. Three research assistants (one Indigenous and two non-Indigenous) helped with the transcription of 80 percent of the interviews. I transcribed 20 percent of the interviews. I used ATLAS .ti to analyze the data and generate forty-five exploratory codes. I then refined the list and came up with eighteen subcodes that helped identify findings and patterns.

## TOURISM-BASED DEVELOPMENT IN INDIGENOUS VILLAGES

In the summer of 2008, I traveled for the first time to the village of El Ranchito. I took a horseback tour with one of my brothers who lives in Copán, a tourist from Switzerland, and the owner of a tourism agency who sold us the tour. We made our way slowly under a light rain on horses that were far too small for us. We traveled for three kilometers until we reached the foot of a steep hill. It had been raining for four days, so the dirt roads were nearly destroyed. At that point, El Ranchito—located atop the hill—was only reachable on foot or by horse. I traveled there to meet an Indigenous family who received funds from the World Bank's *Fondo Prosperidad* (Prosperity Fund) to open up a traditional jewelry shop. The family at El Ranchito were one of three Indigenous families in the region of Copán who received funds from the World Bank for tourism-related development projects.

After passing a few houses on the main road, we reached the family shop. It was a one-story new building that stood apart from the adobe houses in the village. A colorful wooden sign at the entrance read, "Mayan Princesses. This project is financed by Prosperity Fund." Underneath the title was the logo of the Honduran Ministry of Tourism and the World Bank. As we entered the store, we were greeted by a young girl about 10 years of age and three older women. The women were working with different kinds of seeds. The walls of the building were populated by necklaces, earrings, bracelets, and other items created at the shop. There was also a banner hanging from the wall that read, "Ch'orti' Maya Jewelry" and showed images of all of the different kinds of seeds used to make the crafts.

I approached one of the young ladies who was working with an electrical machine to pierce the seeds. We talked about the process of collecting the seeds. She said the family had to travel to nearby villages to collect the seeds or buy them from other Indigenous people. I then talked to the owner of the store—Doña Dolores (pseudonym). Unlike other Indigenous people in the same village, who are often reserved and unwilling to talk to outsiders, Doña Dolores was open, detailed, and unafraid to speak candidly about her experience securing funding to open her store. We stayed for a few hours to converse with everyone. While we ate lunch together, Doña Dolores talked about the World Bank's Prosperity Fund project. She lamented the fact that only three Indigenous families were awarded funds out of the sixty-nine projects funded. The other two initiatives were funded in two Copán villages. One project consisted of a restaurant and a textile shop. The second project funded a ceramic workshop and store.

Similar to Doña Carmen, who I quoted at the start of the chapter, Doña Dolores explained that the majority of the families who benefited from the

World Bank's initiative were middle-class non-Indigenous families who already owned the parcels of land required by the project to establish a tourism-related business. The project required participants to devise a business idea and show proof that they had land to build the proposed project. An external committee would then assess the viability and originality of the project and would award the funds based on these elements. Doña Dolores explained how some non-Indigenous families received up to half a million lempiras (at the time, US$25,000) for their projects. Her voice began to break as she explained how often she cried and prayed to be granted an award. "I suffered many humiliations," Doña Dolores noted with a low voice. I understood that the sense of humiliation she was referring to had to do with her interactions with the evaluation committee comprised in its majority by non-Indigenous people from the urban sector of Honduras who often discriminate against Indigenous people and people (Indigenous and non-Indigenous) from rural areas. "In the end it paid off (*valió la pena*)," she said. That day alone, the tourist who accompanied us on the trip spent close to US$80 on merchandise from her shop. At the time of my visit to El Ranchito, the Prosperity Fund project seemed like a success story, at least for Doña Dolores. The following year (2009), however, with the coup d'état that deposed President Manuel Zelaya, these and other tourism projects encountered major obstacles that made it difficult for them to remain successful despite the transition to democratically elected officials in 2010.

During the years leading up to the coup d'état, various Indigenous villages began to be involved in small tourism projects such as horseback riding tours sold by non-Indigenous tour guides. Prior to the 2000s, horseback riding tours to Indigenous villages were rarely advertised. Encounters between tourists and Indigenous people usually happened when handicraft sellers (the majority of whom were children) would travel to the town of Copán from the village of La Pintada to sell souvenir dolls made out of cornhusk. By 2012, when I returned to Copán to complete my dissertation research, tourism activities among contemporary Indigenous communities had increased significantly. For instance, the number of children selling dolls had more than tripled. Non-Indigenous entrepreneurs put together theatrical performances about the ancient Maya to sell to tourist groups. Local tour guides working at Copán's Archaeological Park started emphasizing more horseback riding tours to a few Indigenous villages, and a local indigenous council secured short-term work for some Indigenous people at the Archaeological Park. Additionally, some Indigenous people began to organize more comprehensive tours of a couple of communities. As a young Indigenous educator explained in 2012, "I love tourism . . . I help connect tour operators [with villages] . . . I do this so that tourists visit [villages like] el Carrizalon and La Pintada . . . I try to promote my culture among tourists." While tourism activities certainly opened up livelihood paths for some Indigenous people within and outside their villages, they also created a competitive market environment (for everyone in the region), making identity a contentious phenomenon. In the next few sections, I explore some of the outcomes stemming from an increase in tourism-based activities.

## SOME CHALLENGES IN TOURISM-BASED INITIATIVES

Since the 1960s, the World Bank has funded tourism-related development work around the world as a strategy to mitigate poverty. In the 1990s, the Honduran government affirmed that in order to combat poverty, tourism was as important as manufacturing and agriculture. Especially the Callejas administration's turn to neoliberal reforms paved the way to supporting the tourism industry. During the same decade, the idea of "sustainable development" was beginning to surface as a strategy to combat economic poverty. Within the framework of "sustainable development," surged the idea of "sustainable tourism." Sustainable tourism comprised both the idea of environmental awareness and "social sustainability." As Hawkins and Mann (2007, 353–357) indicate, during the 1990s "projects with direct or indirect tourism involvement and benefit were mainly oriented toward environmental, cultural, and social themes." By the early 2000s, a focus on sustainable development had turned Honduras into a "'pilot' country for World Bank tourism microenterprise funding" (Ferguson 2010, 5). It was through these tourism microenterprise funds that the Regional Development of Copán Valley's Project (RDCP) was financed from 2003 to 2009. The Prosperity Fund was a subset of RDCP, allocating close to US$3 million to fund different tourism-related projects in Copán (Ferguson 2010, 10–11).

Prior to the Prosperity Fund project, the World Bank sponsored the Interactive Environmental Learning and Science Promotion Project (also known as *Profuturo*). *Profuturo* was the earliest endeavor funded by the World Bank in Copán through a local organization called Copán Association. For many years, Copán Association was managed by Ricardo Agurcia—one of Honduras' most notable archaeologists—with support from the National Pedagogical University of Honduras (UPN). In 1999, *Profuturo* allotted US$8.3 million for sustainable development projects that would support and expand "scientific, environmental and cultural knowledge and management" (Junho Pena & Johns Swartz 2002, 1). Even though most of the *Profuturo* funds were allocated for management initiatives at the Copán Archaeological Park, one segment of the project funded different "Pre-Hispanic Heritage" initiatives. In line with this emphasis on heritage, one of the main goals of the project was to "provide scientific knowledge for teachers and students, [and] promote ownership among Indigenous peoples" (ownership in this context referred to Copán's Archaeological Park). Eighteen Indigenous people from three different villages were selected by the Ch'orti Maya Indigenous Council of Honduras (CONIMCHH). The chosen Indigenous participants received different capacitation workshops emphasizing topics such as the history of Honduras, archaeology, astronomy, biodiversity, communication, English-language classes, museography, and serving as a tour guide (Junho Pena & Johns Swartz 2002, 2). With regard to *Profuturo*, a report from the World Bank assured that:

> As a result of this training program . . . Chortí villages are beginning to make plans for tourism in their villages and for possible employment in the park. . . . The training program was timely and it has served as a catalyst for the aspirations and ideas that the communities have for tourism. By providing knowledge and skills training, it has empowered the communities to provide their own guiding and other services, rather than rely on intermediaries. (Junho Pena & Johns Swartz 2002, 2)

Though a few Indigenous families were inspired by this program to seek funds for tourism-related projects (via *Fondo Prosperidad*) they encountered a myriad of obstacles. As I explained earlier through Doña Dolores' account, because of the requirement to own land, not all Indigenous families qualified for loans. Those families who did qualify for loans and grew their businesses experienced resentment by other families in their villages who felt the loans had increased economic disparities in their villages. For example, by 2016, Doña Dolores could no longer keep her store. In addition to tourism fluctuations, she started receiving death threats in her community as a result of her economic success and because she had altered male-dominated traditional household dynamics. Since Indigenous councils elect new officials every two years, a new rural council (comprised mostly of men) broke into her store and removed all of her handicrafts. She was eventually asked to vacate the natural jewelry store building and was not able to start her business again. In recent informal conversations with Doña Dolores, she explained that none of the Prosperity Funds projects given to indigenous people were able to survive. In her case, she explained,

> the land where the jewelry store is located, belongs to the community and is administered by the councils. Some of the people in the councils did not like to see me work [succeed] so they did everything they could to kick me out.

Even though different families have tried to get help from Indigenous councils, internal issues have left many Indigenous families without support. She explained, "we have tried to get help from the organizations [Indigenous councils] but due to corruption and internal divisions we no longer know which leaders to follow . . . there is nothing left for us there" (informal conversations 2020).

Being able to make a living through tourism-related activities is already a competitive practice in Copán. For Indigenous people, it is even more difficult as they lack the necessary economic and social support (e.g., land in the town; connections with non-Indigenous tourism workers) needed to succeed in the industry. This is an issue for which the World Bank's tourism-development programs have been criticized in the past. Hawkins and Mann (2007, 352) argue that even though the concept of sustainable development encouraged non-governmental organizations and other donors to support tourism projects, numerous studies found a significant gap between the idea of tourism-related sustainable development and the actual opportunities (and benefits) of the practice. *Fondo Prosperidad* has been criticized for the way in which projects were distributed. Similar to Doña Dolores' claims, several residents in Copán (Indigenous and non-Indigenous) have confirmed that these tourism-related projects were awarded to middle-class people who did not need as much economic support as other people in need. Meanwhile, some non-Indigenous entrepreneurs argued that while Indigenous villages have the potential for identity-based tourism projects, Indigenous people do not possess the adequate training and infrastructure to establish and support tourism businesses.

## CULTURE AND TOURISM

James David Wolfensohn, who was president of the World Bank from 1995 to 2005, continually supported the idea that culture was intrinsic to investment and an important conduit for human development. According to Wolfensohn, "There are development dimensions of culture. Physical and expressive culture is an undervalued resource in developing countries. It can earn income, through tourism, crafts, and other cultural enterprises" (Yudice 2003, 13). This notion of culture as a "source of value," explains Anderson (2013), is connected to the neoliberal multiculturalism project embraced by many Central American governments. Neoliberal multiculturalism, understood by some (e.g., Hale 2005) as a governance scheme deployed by multilateral organizations to create and restrict ethnic differences, promotes ethnic diversity through the tourism industry. For Anderson, neoliberal multiculturalism weds cultural difference with the accumulation of capital. Anderson (2013, 277) explains: "By exploring how value is produced and appropriated out of ethnicity, we can also analyze how multicultural projects induce multiple responses from ethnic subjects enticed to participate by the promises of recognition and inclusion." In a context resembling that of the Ch'orti' Maya, Anderson found that in the Garifuna village of Sambo Creek (in the Honduran Caribbean coast), many people lamented that their culture was being used and exploited in the tourism industry by wealthy non-Garifuna entrepreneurs and foreign people living in their village.

Other scholars have proposed that tourism can potentially support Indigenous identity formation. For example, Laurie Medina's (2003) work with the Mopan Maya of Belize shows that Indigenous people have relied on tourism-servicing and the publications of archaeologists as "new channels" for accessing cultural practices no longer accessible to them. Responding to assertions that the commoditization of cultural practices in tourism contexts makes these practices unauthentic, Medina argues that tourism among the Mopan Maya does not lead to the rise of a "new" or "commoditized" culture that is different from a more authentic culture. Instead, tourism has inspired some Mopan Maya to read the works of Western archaeologists and epigraphers in order to learn more about lost cultural elements (e.g., cosmological knowledge and language) (Medina 2003, 357–365).

Similarly, for the region of Copán, Mortensen (2009a) argues that "cultural tourism" constitutes the most salient factor in encouraging the creation of Ch'orti' Maya ethnic revitalization projects. As Mortensen (2009, 251) explains, "by providing a market for cultural distinction, and thereby ascribing it economic value, cultural tourism potentially can provide the means and motivation for reviving cultural traditions, language, and pride." Nevertheless, Mortensen is also critical of some issues of power surrounding cultural representation in the tourism industry in Copán and its effects on the Ch'orti' Maya. Mortensen argues that the homogenous picture of the "Maya" produced at the intersection of tourism and archaeology can undermine the efforts of contemporary Indigenous groups to produce their own cultural renditions (2009a, 247–252). These kinds of conflicts of representation have emerged more notably in recent years with the production of theater performances for tourists.

An ethnically diverse image also benefits the Honduran state as it works toward making the country an attractive multicultural tourist destination and also a good candidate for funding development agencies. Cultural differences figure prominently in tourism-related business models. As Anderson (2013, 278) indicates, funding agencies work with the assumption that "local ethnic groups can participate in tourist markets—beyond the role of workers—to create business, sell goods and services, and take advantage of their cultural distinctiveness." In the case of Copán, while it is the case that some Indigenous families have benefited from tourism-related activities, the business of ethnic diversity is something that has mostly been exploited by non-Indigenous society. An event that undressed some of these issues of representation was the December 21st (2012) celebration of the end of the thirteenth Baktun in the Mesoamerican long count calendar, popularly known popularly as the "Mayan apocalypse." This celebration, which was held in Copán as well as other sites in Mexico and Guatemala, revealed some of the conflicts stemming from the politics of representation in heritage-based tourism.

By 2012, tourism-related activities had become more common among Indigenous people as a livelihood approach. However, similar to other commercial opportunities in Copán, making money off of tourism became a controversial issue among Indigenous people themselves and other non-Indigenous people involved in tourism-servicing. As I will explain later in this chapter, when Ch'orti' Maya cultural elements became a lucrative idea for the tourism market, non-Indigenous people with access to both economic and social capital quickly capitalized on some of these elements and reworked them into business ideas. Simultaneously, some Indigenous people turned to activism in order to gain support and compete with non-Indigenous entrepreneurs in spaces such as ritual performances. The gap between Indigenous mobilization to achieve cultural goals and Indigenous mobilization to secure livelihood opportunities has been discussed in the literature on tourism and Maya identity. For the region of Chiapas in Southern Mexico, Van de Berghe (1995, 576–583) shows how Indigenous mobilization has emphasized goals such as land recovery and sociopolitical autonomy over economic opportunities created by identity-based tourism. Similar to the case of Copán, Van den Berghe explains that in Chiapas, tourism development has helped turn identity into a "marketable resource." However, because of their "political, economic, locational, and linguistic advantages," it is non-Indigenous elites who end up reaping the economic benefits of the tourism market (Van de Berghe 1995, 576–583).

For the area of Guatemala, which is home to close to twenty-one Maya Indigenous groups, the literature on Maya activism has understated the complex interaction between livelihood opportunities and political identity (Little 2004b, 43). In his work among the Kaqchikel Maya of this region, Little (2003, 2004a, b) contends that even though Pan-Maya Indigenous leaders and activists have worked toward their cultural goals and reinforced their identity through their activism, other Indigenous people choose to deliberately mobilize (and produce) their particular identities in the marketplace. By performing their identities in the marketplace, some Indigenous people evade having to commit to political ideals that could compromise their position as vendors. Securing a permit to sell handicrafts in marketplaces is a highly political and sometimes lengthy process (Carey 2008; Little 2004a; 8–9, 18, 2004b, 43). In Copán, the case of the Ch'orti' Maya reveals a combined strategy wherein Indigenous people mobilize their identity in both their activism and tourism-related activities. CONIMCHH (which eventually split into two different Indigenous councils) represents the first organizing assembly for Indigenous people in Copán. Its establishment in the early 1990s allowed Indigenous villages to come together as Ch'orti' Maya, demand land and ethnic recognition from the Honduran government, and receive international support from transnational activists (Rodriguez-Mejia 2016). As I explain in the following section, the ability of the Ch'orti' Maya to inhabit both the activism and tourism realms, especially through the support of CONIMCHH, allows them to contest non-Indigenous representations of their culture for commercial purposes.

## CELEBRATING THE MAYA APOCALYPSE

During the month of December 2012, thousands of tourists from all over the world filled the streets of the small town of Copán. There were so many visitors that hotels quickly ran out of rooms to host people. Some residents began to rent out their backyards for people to camp out. The crowds had arrived to be close to a ceremonial site like Copán to commemorate the end of the thirteenth *b'aktun* or last cycle of the Maya Long Count calendar to take place on December 21st (Restall & Solari 2011). Several entities, including the Honduran government, the Chamber of Commerce of Copán, non-Indigenous business people, and even foreign filmmakers, were invested in the week-long festivities taking place prior to December 21st—which was advertised internationally as the "end of the world." To honor the day, government officials, along with local, national, and international tour businesses, supported specialized tours of the archaeological park of Copán. The park, which is a UNESCO-recognized World Heritage site, is a well-known Classic Maya ceremonial center, and hence it stood as both the main attraction and the basis for the events that took place. But the celebration went beyond the festivities at the park. Government agencies also organized VIP celebrations, non-Indigenous business people assembled ornate theatrical shows portraying Classic Maya rituals for tourists, and a local planning board hosted exhibit openings, movie screenings, and concerts by national singers.

On a different part of the park, CONIMCHH organized a more reserved ritual that stood apart from the overelaborate celebrations that took place in the town. In attendance were a handful of members of the council, a few outsiders, and news reporters. The solemn tone of the event served as a crucial counterpoint to the other celebrations seeking to take advantage of people's fascination with the ancient Maya. The enormous display of promotional celebrations was prepared by PLAN 2012—a planning board comprised of local non-Indigenous scholars and businesspeople. PLAN 2012 began organizing events as early as March of 2012. Some of these events included presentations by prominent archaeologists, symposiums, and visits to several museums. And while their printed program specified that the goal of these activities was to "remember" the traditions of the ancient Maya, the board did not plan any event involving the contemporary Ch'orti' Maya. This was to be expected as the relationships between the Ch'orti' Maya, the Honduran government, local non-Indigenous leaders, and the archaeological park itself have been historically contentious. In fact, almost every year since 1998, the Ch'orti' Maya have organized strikes at the Copán archaeological park, sometimes blocking access to the facilities in order to negotiate land transfers with the Honduran government. Protests are usually followed by negotiations with state representatives, but in the last twenty years, the state has only managed to purchase and transfer small parcels of land and secure a few provisional jobs for Indigenous people at the Archaeological Park and local museums.

The omission of the Ch'orti' Maya from the December 2012 events program remains somewhat surprising, especially because since the mid-1990s the idea of "ethnic difference" (especially Maya indigeneity) began to be considered "added-value" (Anderson 2013) in the realm of cultural tourism. The tourism industry actually helped increase the interaction between the Ch'orti' Maya and local non-Indigenous entrepreneurs and politicians, leading in some instances to unexpected partnerships. For example, as I previously explained, tour operators as well as tour guides in the town of Copán began to arrange tours to some of the few Ch'orti' Maya villages that had tourism-related establishments. On the political realm, the Honduran government—especially during the Porfirio Lobo Sosa administration—began to intentionally encourage the integration of Indigenous society in the tourism industry. For instance, in August of 2012, the Ch'orti' Maya of Copán hosted the commemoration of the International Day of the World's Indigenous Peoples at Copán' Archaeological Park. Hundreds of Indigenous people from different parts of Honduras joined the Ch'orti' Maya at the event. The celebration was also attended by state officials from other Central American countries, delegates from the United Nations, and Porfirio Lobo Sosa. During the event, despite the indignation of many non-Indigenous residents of Copán, Lobo Sosa recommended that the name Copán Ruinas (which is the official name of the municipal region) should be replaced by the name Copán Galel (the last Ch'orti' Maya leader in the 1500s). Lobo Sosa's public affirmation of support for the Indigenous sector (especially in the tourism industry) and the presence of other political figures, signaled a closer connection between the state and the Ch'orti' Maya. What began with the inclusion of a few Indigenous families in *Fondo Prosperidad* seemed, at the time, to have grown into a more concrete plan of integrating the Ch'orti' Maya into various aspects of the tourism market. However, the peripheral presence of the Ch'orti' Maya in the December 2012 celebrations revealed deeper social conflicts outside the political arena. The exclusionary tactics employed by PLAN 2012 prevented the Ch'orti' Maya from playing a more central role in the different events. Yet, as I explain in the next section, the December 2012 events pushed the Ch'orti' Maya to rely on their activism in order to contest their marginalization and increase their cultural visibility through other mediums.

## A CH'ORTI' MAYA CEREMONY

In the cold, pre-dawn hours of December 21, 2012, I waited by the entrance of *Las Sepulturas*—a small archaeological site two kilometers east of Copán—with anthropologists Lena Mortensen and Adalid Martínez Perdomo. Two reporters, one from CNN and another from a local news station, waited with us for the arrival of Indigenous leaders from CONIMCHH. These Indigenous leaders, including a shaman, were there to carry out a ritual honoring the end of the thirteenth Baktun. The reporters seemed excited as more Indigenous people began to arrive wearing traditional Indigenous attire. A few more reporters also joined the ceremony, causing a bit of commotion as they made their way through the crowd with their heavy equipment. Once the ritual was about to begin, the shaman asked the attendees to refrain from taking photographs or videos. Everyone except two cameramen followed the shaman's request. Members of the crowd began to reprimand the two cameramen who continued to film and take pictures.

Though this ceremony was covered by the media, it was excluded from the main list of events prepared by PLAN 2012 for the December 21st celebrations. The planning committee organized and heavily advertised their own opening and closing ceremonies for which they hired their own Indigenous shamans. Not only did PLAN 2012 secure the main archaeological site for their ceremonies, but they also opened them to the larger public. The Ch'orti' Maya from CONIMCHH had no choice but to perform their ceremony at *Las Sepulturas*. Despite their exclusion from the advertised events, the Ch'orti' Maya used their political connections through CONIMCHH to invite national and international media reporters, scholars, and other non-Indigenous activists in support of their cause. We had all assembled at *Las Sepulturas* by 4 a.m., waiting for the ritual to start at dawn. Performing the ceremony at this specific site and at this particular time allowed the Ch'orti' Maya to set themselves apart from the more commercial ceremonies that were attended by thousands of tourists and the highly staged events targeting exclusive audiences. The liminal position of *Las Sepulturas*—located on the edges of the larger archaeological park of Copán, but still within the borders of the protected area—added legitimacy to the event despite its exclusion from the broader, government-sponsored program. In effect, being able to conduct their ceremony in a separate space enabled members of CONIMCHH to transform a typical act of marginalization into a mediated opportunity to affirm their legitimacy as an ethnic group. At the conclusion of the ceremony, news reporters were once again eager to interview all of the participants. But this time, Indigenous leaders took the opportunity to strategically bring up topics such as Indigenous rights over the Maya ruins, ancestral land, and their own traditions. In this sense, they turned non-Indigenous society's exclusionary acts into a way to advance their own activism goals. (See figure 6.1.)

A week after the December 21st festivities had ended, I ran into Doña Candida Garcia (pseudonym), one of two shamans who were hired by PLAN 2012 to perform their opening ceremony. She seemed tired as a result of having to rehearse with PLAN 2012 for several days leading up to the events. For all of the days she spent rehearsing and performing the performance of the ceremonies the committee paid her 500 lempiras (US$20). She was hired by José Cordero (pseudonym), one of the most active non-Indigenous

**Figure 6.1 One of the Ceremonies Sponsored by PLAN 2012.** *Source*: Photo by the author.

entrepreneurs involved with PLAN 2012 and the producer of theater performances involving ancient Maya stories. Since Cordero had previously worked with Doña Candida before, he was able to convince her to participate in the state-sponsored festivities even though she was an active member of CONIMCHH. "They . . . put me in a difficult position with CONIMCHH," she said, because of how the committee excluded the Indigenous council from the planning stages of the events. But Doña Candida accepted the invitation because she needed the money.

The same week, while I was working on my field notes at a local café, Cordero came to talk to me. He seemed agitated. As a member of PLAN 2012, he was heavily invested in organizing many of the events that took place the week of December 21st. When I asked how the events went, his voice escalated. He said he was upset with Doña Carmen Mendoza (pseudonym), who was the other shaman hired to perform a closing ceremony. Cordero lamented that when Doña Carmen had concluded her ceremony, she began to speak to the crowd about Indigenous rights and the rights of their organizations to manage the archaeological site of Copán. He pulled her aside and scolded her, telling her that those kinds of discussions did not have a place in that context and that she should know when to bring them up. Seven months later, I had dinner with Cordero. He talked about the possibility of organizing a "Maya" ceremony at the official opening of a small archaeological site near *Las Sepulturas*. Cordero made the case that ceremonies such as those would bring Indigenous people closer to their patrimony. I remembered our conversation seven months prior and asked him if he thought that the December 21st events brought Indigenous people closer to their patrimony. He became serious and paused for a while. He argued that CONIMCHH's exclusion from PLAN 2012 was the former's fault for refusing to come up with a ceremony that would work according to the committee's plan.

Cordero's case is intriguing in the sense that despite his adversarial relationship with some Ch'orti' Maya activists, he openly supports Indigenous activism and criticizes non-Indigenous elites for the mistreatment of many Indigenous people over the years. In one of our interviews, Cordero mentioned that although he is not considered Indigenous, he wanted to send blood samples to National Geographic to determine what percentage of his genes were Maya. Moreover, Cordero often criticized the dominant presence of Western archaeologists in Copán, yet his theater performances draw from the publications of some foreign archaeologists.

Although Cordero conducted his undergraduate studies in the United States and his graduate work in Europe, he argued that his professional goals have always revolved around helping rural and Indigenous groups in Copán. When I asked him what the word Ch'orti' Maya meant to him, he argued that it meant his ancestors and a living culture. Since he was in elementary school, he said, teachers had taught him that Maya culture was dead, but that to him, it represented a living culture and a culture that "lives inside our genes." Even though, to most people, Cordero would not be considered Indigenous, he actually argued that the only difference between him and Ch'orti' Maya people is the fact that he was born in Copán and they were born in the villages.

Cordero also argued that there should not be a difference between the term ancient Maya and Maya Ch'orti'; his training of Indigenous people to participate in performances locally and outside of Copán, he argued, only contributes to empowering many of them who still feel shy about saying that they are Maya. Cordero further argued that his projects contributed to the visibility of Indigenous people while providing an avenue for them to "connect with their ancestors." Independently of whether or not people feel connected to their ancestors by participating in theater performances, Cordero's projects have played an important role in integrating some Ch'orti' Maya people into the tourism industry. In effect, prior to Cordero's initiatives, the word Maya in the tourism realm was, for the most part, associated with Copán Archaeological Park and its history. The next section addresses some of the subjective understandings of identity-based tourism initiatives in some of Cordero's projects.

## THEATER PERFORMANCES AT THE INTERSECTION OF TOURISM AND ACTIVISM

Cordero began producing theater performances since the early 2010s. His theater production plays two different roles: on the one hand, they portray the Ch'orti' Maya as timeless and exotic figures that appeal to foreign tourists and are consumed as "museum pieces" (Mortensen 2009). On the other hand, they make the existence of the Ch'orti' Maya known to many people who previously had only associated the word "Maya" with the archaeological site of Copán. For instance, a few months prior to the December 21st celebrations, Cordero was hired to take Indigenous actors to San Pedro Sula (the second biggest city in Honduras) to the premiere of the film *2012: The Beginning*. During the premiere, which took place in the theaters of one of the city's biggest shopping malls, Cordero prepared a performance in which Indigenous people, dressed as (he imagined) people would have during the Classic Maya era, walked around while people took pictures with them and asked them questions about their lives and communities. On a previous conversation with Cordero, he pointed out that one of his dreams was to be able to have a group of Indigenous actors walk around the Copán Archaeological Park dressed in "Classic Maya" costumes so tourists could interact with them and ask questions about their communities. According to Cordero, these practices bring

Indigenous people closer to the distant Maya past that is known to tourists. As he put it,

> being involved in these projects helps them believe more in their culture and recognize their own identity and to feel proud about the fact that the Maya were not those mercenaries that we see in the movies . . . but that they were artists, they knew the art of dancing, the art of loving nature.

For example, Cordero also claims to have taught Indigenous people to be comfortable playing the role of a shaman or medicine person. In Copán, the idea of a shaman is often associated with evil and sorcery; thus, Cordero argued that it was important for Indigenous people to develop a different understanding of this concept and accept it as an important part of their culture.

While the actors in Cordero's group (close to fifty people) were mostly people from Indigenous villages affiliated with CONIMCHH, the organization never interfered with or complained about Cordero's business ventures. One of the reasons is that Cordero, with the help of other aficionados of ancient Maya culture, had created performances strictly about the ancient Maya that did not borrow any elements of contemporary Ch'orti' Maya culture. In fact, the only connection between Cordero's performances and the present Ch'orti' Maya were the actors themselves, whom he hired for their physical appearance. Community leaders were upset that these projects were managed by people who they considered non-Ch'orti'-Maya, but as Cordero explained, people in the communities do not have the resources to compete and be able to form their own groups. It turns out that this was not the case. A few months after talking to Cordero, one Indigenous leader decided to take action. The fact that many of the actors belonged to communities affiliated with CONIMCHH inspired the leader to convince them to form their own group and become affiliated with CONIMCHH as a theater group.

While conversing with the Indigenous leader who led the split of the theater group, he explained that he thought the actors were not compensated as well as they should be and that it was fair that they had an Indigenous leader. Cordero, on the other hand, argued that the previous manager of the theater group (also a non-Indigenous resident of Copán) paid them barely any money for their performances. Cordero explains that when he began managing the group, he asked the participants how much money they wanted to earn per performance because he wanted them to be well compensated, but some actors have contested these claims. Most participants still abandoned the group (at least temporarily). By 2012, the new group had begun applying for grants through the Ministry of the Indigenous and Afro-descendant Groups of Honduras (SEDINAFROH) to become established as the official theater group of CONIMCHH. They called themselves *Grupo Ajaw* (lord). As of September 2012, the *Grupo Ajaw* had also started delivering their first performance in an Indigenous village (see figure 6.2) as part of the *Festival del Maiz* agrarian ceremony, which is celebrated yearly in several villages in order to bless their *milpas* (an Indigenous agricultural technique involving the plantation of corn, beans, and squash within the same parcel of land).

One important detail to note is that while *Grupo Ajaw* did intend to perform during tourism events, this particular performance was done in order to support CONIMCHH and the San Rafael community's efforts to promote Ch'orti' Maya identity. In fact, the vision of the group's leader was to make these performances a permanent element of the *Festival del Elote* and increase the visibility of their communities for both tourism and government officials. The separation of the actors from the first theater group denotes another strategy of contestation made possible by the support of CONIMCHH. Cordero recognized that Indigenous people did not always trust working with him for fear of being exploited (as the previous non-Indigenous manager did). However, he claims to have explained to the actors that they were free to perform for their communities in their own free time. Cordero explains that the level of distrust from Indigenous people toward him stems from a larger sense of distrust toward non-Indigenous society in Copán following years of racism and discrimination.

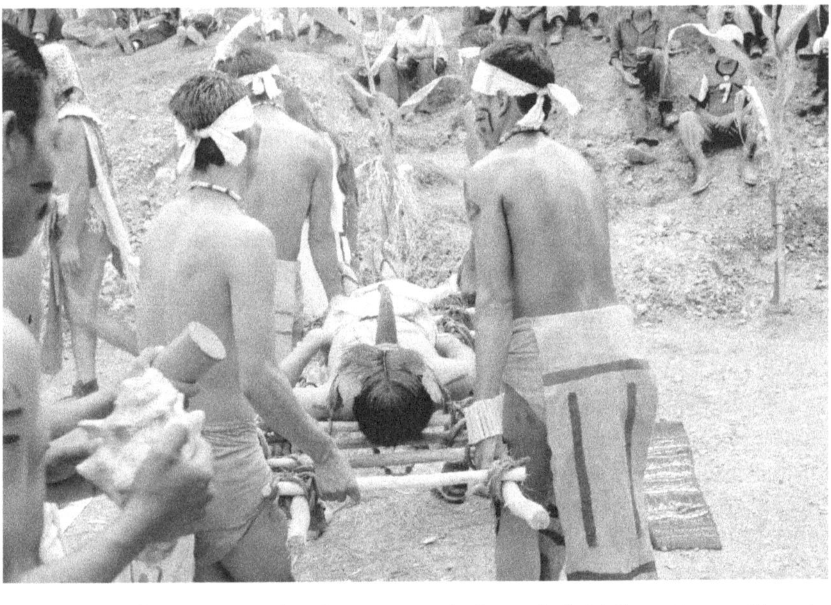

**Figure 6.2 Ajaw Group Performing at San Rafael's Festival del Elote.** *Source*: Photo by the author.

Theater performances are one of the most lucrative identity-based tourism livelihood strategies that have surged at Copán. One of the group members noted that some foreign tourist groups would pay up to US$2,000 per performance. This would explain why these shows rapidly became a commodity. While the *Ajaw* began to compete with a separate group organized by Cordero, the latter competed with other theater groups that originated in other (non-Maya) regions of Honduras. For example, during the December 21st festivities, PLAN 2012 hired another theater group to deliver a performance during the day, while Cordero's group delivered a performance in the evening (see figures 6.3 and 6.4).

Theater performances are also a difficult livelihood strategy to participate in since they require a significant investment for costumes, makeup, traveling expenses, and renting performance spaces that the average Indigenous family would not be able to afford. In this sense, the Indigenous performance group is mostly beneficial to activism in helping give legitimacy to their communities and also to the Ch'orti' Maya as a group who now also use the theater group when government officials come to visit Copán. Other, less lucrative strategies, as the following section explores, make use of contemporary Ch'orti' Maya identity in more creative ways.

**Figure 6.3 Theater Group from Copán Depicts a Sacrifice Ritual in Honor of Copan's Dynasty Founder K'inich Yax K'uk' Mo'.** *Source*: Photo by the author.

**Figure 6.4 Theater Group from Outside of Copán Performing a Depiction from the Popol Vuh Maya Creation Myth.** *Source*: Photo by the author.

## LAS MUÑECAS DE MAIZ

*Las Muñecas de Maiz*, or corn-husk dolls, is an expression that has become famous in Copán to refer to the children who walk to the town from the village of La Pintada to sell these handicrafts to national and foreign tourists. Occasionally, it is difficult for them to distinguish tourists from locals. At times, some of the locals would say, "I am not a tourist," in order to preclude the children from asking them to buy the handicrafts. With tourists, however, the young vendors are relentless. If a tourist ignores them, they continue to follow them—sometimes even making friends with them. The initiative for the corn-husk dolls surged in the late 1990s during the founding years of Ch'orti' Maya activism at Copán and the establishment of the first Indigenous council. Some development non-governmental organizations facilitated workshops that would teach Indigenous people identity-based livelihood strategies. Consequently, one of the projects focused on the production of these corn-husk dolls together with textiles. As I mentioned previously, in 2008 there were only a handful of children selling these dolls. In 2012, the number of children selling dolls had more than tripled; more families were producing in the village of La Pintada and sending their children to town to sell them.

For some of these families, selling these dolls is the only form of subsistence. As one Indigenous leader explained about the village of La Pintada, "[selling dolls] has been very useful for families because the men work in their small parcels of land and only make a little money and with that little money they can only provide a small fraction to buy food." Because of the difficult economic situation, the children are very persistent. If tourists refuse to buy their handicrafts, the children usually offer to sing the Honduran national anthem in Ch'orti' Maya or recite the Our Father. These are two things they have learned and memorized from their elementary school teachers. For the region of Copán, as mandated by the Ministry of Education, Indigenous as well as non-Indigenous teachers (sometimes against their will) are required to teach the Ch'orti' Maya language from elementary school to high school. This initiative derives from the continent-wide Intercultural Bilingual Education (IBE) project, which deals, in part, with the survival of Indigenous languages across Latin America using the education system as an avenue (Lopez & Sichra, 2007; Gustafson, 2009). Although it has been a difficult program to implement, in the sense that the teachers are not fluent Ch'orti' Maya speakers, they have ensured that the children learn the Honduran National Anthem and the Our Father. Thus, it is not surprising to see the children creatively put these cultural elements to use while asserting their identity for economic purposes.

In the tourist's eyes, and even to Honduras' non-Indigenous society, the selling of corn-husk dolls is a symbol of identity that constructs the Ch'orti' Maya as a unified ethnic group with at least some usage of their native language. I argue that one of the reasons parents and activists have encouraged children to sing the Honduran National Anthem and recite the Our Father is to both assert the identity of their communities (who are mostly Catholic) and contest the criticism of non-Indigenous Copán society that the Ch'orti' Maya are Guatemalan. By singing the Honduran national anthem in Ch'orti' Maya they are able to also assert their identity as both Indigenous and Hondurans. Using a play on words by Dario Euraque (2002), who writes about the Mayanization of other Indigenous groups in Honduras, I view this practice as the *Hondurization* of the Ch'orti' Maya, who for years have been criticized as being Guatemalan due to the strong family ties that they maintain across the border (Metz 2009; Metz et al. 2009; Rodriguez-Mejia 2016). (See figure 6.5.)

While in Copán, the children who sell the dolls target specific areas or restaurants in town, in the communities, families compete to produce these handicrafts. One Indigenous woman who works with dolls and textiles explains, for instance, that tour guides only take tourists to buy their products in one specific part of her community, forcing those who do not live in that section to send their children on a forty-five-minute walk to sell the products in the town of Copán. In March (2013), the first time I traveled to the community of La Pintada to meet with their rural Indigenous council, the

**Figure 6.5 Indigenous Children from the Community of La Pintada Selling Dolls.**
*Source*: Photo by the author.

mototaxi would only take me to the section of the village where the school, the community center, and the so-called *casa de artesanias* (artisans' house) are located. There I was greeted by several children carrying the same dolls. I had to quickly tell them I was not a tourist. I told them I was looking for the *Consejero Mayor* (head) of the community. A young girl said, "you must be looking for my uncle Carlos (pseudonym). He lives on the other side of the village." She pointed far into the distance. I walked for twenty minutes and arrived at the other side of the community, where tourists barely visit. Several of the children who sell dolls in Copán live on that side of the community. In addition to the dolls, textiles, and natural jewelry, another family tried to open a restaurant with funds from the World Bank's *Fondo Prosperidad* project, but it did not work. The third project, in the community of El Carrizalon, obtained a grant to build a ceramic shop. This shop employs other people from the community, and it is owned by the woman who served as the first president of CONIMCHH. While this is perhaps the project that has yielded the best artisan products out of the projects sponsored by the World Bank, the community is so far from the town of Copán (bordering Guatemala) that cars or mototaxis can only take people to a certain point before they have to walk for half hour to reach the community. Thus, although the community is regarded by many Indigenous people as the "most traditional" out of all of the communities, due to having several people there who still speak the Ch'orti' Maya language, not many tourists visit this village. Additionally, the ceramic artifacts are too fragile to be mass-transported to the town of Copán; hence the business has not been lucrative.

## DISCUSSION AND IMPLICATIONS

The role and involvement of the Ch'orti' Maya in the tourism industry of Copán reveal some of the limitations and inadequacies of development initiatives with a focus on tourism and identity. One of the obvious points has to do with the sustainable economic element emphasized in these approaches. As Hawkins and Mann (2007:359) pointed out, tourism has been deemed an "unstable and volatile [industry] . . . with destinations at the mercy of trends and fashions for their popularity, dependent upon fluctuating political and economic conditions worldwide." The Ch'orti' Maya, for example, suffered a major setback, economically and politically, during the ousting of President Manuel Zelaya. A large number of tourism projects sponsored by the World Bank's *Fondo Prosperidad* initiative actually collapsed during the political crisis propelled by the 2009 Coup d'état. Likewise, the outreach work of the Honduran Institute of Anthropology and History (IHAH) under Dario Euraque's administration was discontinued. Prior to the *coup*, the IHAH was implementing projects with Ch'orti' Maya villages to increase their involvement with the Copán Archaeological Park.

During the Lobo Sosa administration, the Ch'orti' Maya received some support related to tourism with the creation of SEDINAFROH (Ministry for Indigenous and Afro-descendant Peoples) in 2009. The role of SEDINAFROH (presently called DINAFRO) has been to appoint representatives chosen from the nine Indigenous and Afro-descendant groups in Honduras to serve as liaisons between the groups and the state. In an interview with Kenia Ramirez, a representative of SEDINAFROH, she argued that this unit was born after the Coup d'état in order to help President Lobo Sosa consolidate Indigenous efforts that had been disrupted as a result of Zelaya's ousting. The establishment of SEDINAFROH also helped to restore trust between Indigenous leaders and the government. During the early 2010s, SEDINAFROH was active in helping bring Ch'orti' Maya demands to the government as well as planning development workshops. The first set of workshops took place in 2012 (thirty-two workshops were facilitated nationwide). I attended one of the workshops taught among the Ch'orti' Maya. This workshop, which was titled, "History, Culture, and Sustainable Development with Identity," lasted three days and was taught by consultants of a private agency from Tegucigalpa to close to fifty Indigenous leaders from different Ch'orti' Maya communities chosen by CONIMCHH. It covered four different units: (1) The origins and history of Indigenous and Afro-descendant peoples of Honduras, (2) Identity and culture among Indigenous and Afro-Honduran peoples, (3) The rights of Indigenous and Afro-Honduran peoples, and (4) the concept of "Buen Vivir (to live well)" and sustainable development with identity. The last unit focused on teaching Indigenous people how to write grants for both national and international granting sources, strategically using their positionality as Indigenous. Another unit emphasized identity formation or, as one of the facilitators explained, "teaching people how to be more Maya." Often throughout the workshop, for instance, when any of the participants would refer to the ancient Maya as "the Maya," a facilitator would quickly correct them and say, "you need to say 'us Maya'." Another aspect that took priority was teaching people the connection between the contemporary Ch'orti' Maya, the Archaeological Park, and the ancient Maya.

These kinds of workshops, as Kenia Ramirez explained, help people think about who they are—"Am I Garifuna? Am I Ch'orti' Maya? Am I Lenca"? However, she also argued that while these initiatives help consolidate the image of Honduras as a pluri-ethnic and multicultural nation, they encourage non-Indigenous *mestizos* (people of mixed ethnicity and race) to also want to belong somewhere. In turn, *mestizos* start claiming that they are Ch'orti' Maya or Lenca, especially politicians trying to win the Indigenous vote. These state-sponsored workshops and tourism development projects underscore the sense of contention that exists in issues of identity. Especially the launching of tourism development projects has created a sense of ethnic reflection whereby people begin to consider the value of identity but also begin to compete for the opportunity to use identity as a source of income. Cases such as the *Fondo Prosperidad*, where 87 percent of the funds were allocated to middle-class non-Indigenous families, exemplify this contention. As a project whose third main objective was to "Rescue, preserve, and promote the cultural, archaeological, and ethnic identity of region" (Rivera 2013), it is surprising that so little emphasis was placed on the Indigenous sector. Yet, on the cover of one of their final reports, the only projects that are showcased through the pictures are those that benefited the few Indigenous families that were part of the initiative. Similarly, the popularity among tourists of performances depicting ancient Maya events has increased the demand as well as competition over the creation of theater groups in and outside of Copán. In the case of Cordero, who has been criticized by some Indigenous leaders for "selling" Maya culture through these performances and using Maya actors, he defends himself by arguing that "we are all Indigenous."

Claims to ownership and issues of competition for job opportunities also figure prominently in the tourism industry in the context of archaeology. From 2006 to 2009, the Honduran Institute of Anthropology & History (IHAH) signed an agreement with the Ch'orti' Maya establishing support to train Indigenous leaders in matters of cultural resources such as cultural heritage and archaeology. IHAH even trained people to collect oral histories in their communities and hired Ch'orti' Maya linguists from Guatemala to travel to Copán's Indigenous villages and teach language lessons. All these projects were also discontinued in 2009 following the *coup*. By 2012, the state had started involving the Indigenous sector again in the activities at the archaeological site. In an interview with Salvador Varela, who was the head of the Copán Archaeological Park that year, he claimed that archaeology was intrinsic for the economic development of Indigenous people. Varela spoke for the potential to open small archaeological sites and museums in Indigenous communities as well as the promotion of living cultures so tourists can visit communities. While Varela is an avid supporter of Ch'orti' Maya activism and has worked closely with CONIMCHH to open twelve rotating posts at the site for Indigenous people, he also believes that in Copán all people have Indigenous blood and should identify as Ch'orti' Maya. Moreover, Varela argued that thirty years of archaeological research in Copán have found that Copán's Indigenous population was comprised of a Lenca/Ch'orti' Maya mixture, and hence identity claims are challenging. This issue is even more problematic considering that since its inception, the Honduran state has promoted the Copán Archaeological Park as one of Honduras' most valuable sources of national identity and heritage (Euraque 2002). Thus, it is not surprising that most residents of the town of Copán claim ownership of the Maya ruins. While state-sponsored initiatives (e.g., SEDINAFROH) and internationally funded tourism development projects (e.g., ProFuturo) inculcated among Indigenous people a sense of ownership over the Maya ruins, the promotion of the park as a national patrimony built by the ancestors of all Hondurans encourages everyone to not only feel a sense of pride over the ruins but also contest Indigenous claims.

## IDENTITY-BASED LIVELIHOOD STRATEGIES IN THE PRESENT

By early 2020, before the COVID-19 pandemic, tourism-related activities in Copán had expanded in creative ways, further revealing the intricacies of identity-based livelihood strategies. Six current developments are worth highlighting:

1. Families from the village of La Pintada continue to send their children to Copán to sell corn-husk dolls. However, these children, some of whom are now accompanied by their mothers, have been spending more time at the Copán Archaeological Park in order to increase their chances of selling their handicrafts.
2. Cordero realized his dream of recreating a Ch'orti' Maya traditional village house and also hiring Indigenous people to hang out at the ruins dressed in traditional attire from the ancient Maya. These Ch'orti' Maya youth charge a small fee for tourists to be able to take pictures with them and they usually hang out on the site on the weekends. Furthermore, Cordero regained control of the *Ajaw* Theater Group for the community of San Rafael because, as Cordero predicted, there was not enough economic and leadership support for this group in their own community. In August 2019, Copán was declared the capital of Cacao of Honduras. For the occasion, and with the support of the ministry of tourism, Cordero organized the first recreation of a traditional Mayan village. He built a traditional hut with an Indigenous family inside (a mother and her two sons). Cordero recreated the same setting at the Copán Archaeological Park (see figures 6.6 and 6.7).
3. The creativeness of wearing traditional items has also been adopted by non-Indigenous people. For example, one popular non-Indigenous

**Figure 6.6 Recreation of a Ch'orti' Maya Household at the Copán Archaeological Park.**
*Source*: Photo taken by Londin Velázquez Martínez.

**Figure 6.7 Indigenous Child Poses for a Photo in the Recreation of a Ch'orti' Maya Household at the Copán Archaeological Park.** *Source*: Photo taken by Londin Velázquez Martínez.

tour guide, Carlos Valdivia (pseudonym), created a headdress using snake skin and macaw feathers in order to attract tourists to hire him for guided tours of the site. He claims the headdress works in his favor as tourists prefer to receive the guided tours from him.

4. Efforts by French Archeologist René Viel, who has lived in Copán for over twety-five years, formed a group of tour guides at the Archaeological Park who call themselves *Guias Unidos*. Four out of ten new tour guides are Ch'orti' Maya. Viel secured both training and work for them as tour guides at the archaeological park.
5. With the help of a local activist and a Christian non-governmental organization, an Indigenous family has begun selling bike riding tours to the village of La Pintada. Because the family is well connected in the village, they can take tourists to talk to an elderly woman who still speaks Ch'orti' Maya, to visit the place where traditional textiles are manufactured, and the homes where the corn-husk dolls are produced.
6. During the presidency of Juan Orlando Hernández, since 2014, there has been a significant shift in tourism operations around the country. As Yobani Peraza, who is the president of the Copan Guide Association in Copán, explains, the government has invested heavily in other regions around the country that can offer tours of colonial towns, cloud forest expeditions, and Cacao plantation tours around Yojoa Lake among others. This redistribution of touristic attractions has caused a decrease in national tourism for Copán as well as visitors from the United States. However, visitors from Europe to Copán are on the rise.

## CONCLUSIONS

The neoliberal era in Honduras, which started during the Callejas administration, opened up the path of sustainable development initiatives, betting on the tourism industry as a valuable asset. These initiatives took place in conjunction with Honduras' re-branding of itself as a multicultural nation. Such re-branding gave the Indigenous sector a chance to emerge as important actors in the new demographic makeup of the country and thus be able to set multiple demands. For the Ch'orti' Maya, economic demands (e.g., land and education) have been minimally met, while the state has focused on increasing the cultural capital of Indigenous communities (e.g., language training, cultural knowledge, identity formation). This has occurred strategically to play a role in the tourism industry. As Anderson (2013, 277–78) explains

> The tourist industry attempts to profit from culture not simply (or even primarily) by churning out cultural commodities (handicrafts, performances, food, and so on), but also by using images of ethnic difference to bolster the image of Honduras as an attractive destination.

An economically impoverished yet culturally diverse destination makes Honduras an ideal candidate for sustainable development funding with a focus on tourism, and the "economic rhetoric espoused by powerful actors promises that the tourist industry will benefit marginalized peoples" (Anderson 2013:278). However, as the case of the Ch'orti' Maya reveals, tourism-based development initiatives have only helped increase the channels through which identity has become commoditized. Left to the forces of the market, Maya identity in Copán has been mostly exploited by those with greater economic, political, and social capital. In this scenario, Indigenous people have resorted to their activism channels in order to be able to compete in a market that has commoditized their identity.

The events of December 21, 2012, the formation of Indigenous theater groups, identity-formation workshops, and the use of cultural elements to maximize the sale of handicrafts are all examples of how identity-related strategies have been used by Indigenous and non-Indigenous people to increase livelihood opportunities. As the Indigenous councils in Copán undergo some internal conflicts, Indigenous people are not able to rely on activism to support their involvement in the tourism industry. However, Ch'orti' Maya students continue to graduate from universities, the number of Indigenous tour guides at the Archaeological Park continues to increase, and some families have begun to organize their own community tours. It remains to be seen how the tourism industry in Copán and Honduras recovers from its near collapse following the COVID-19 pandemic and the role that identity-based tourism and activism play in such recovery.

## REFERENCES

Anderson, Mark. "Notes on Tourism, Ethnicity, and the Politics of Cultural Value in Honduras." In *Central America in the New Millennium: Living Transition and Reimagining Democracy*, edited by Jennifer Burrell and Ellen Moodie, 276–292. New York: Berghahn Books, 2013

Andrews, Wyllys and William L. Fash. *Copán: The History of an Ancient Maya Kingdom.* Santa Fe, NM: School of American Research; Oxford: James Currey, 2005.

Barahona, Marvin. "Del Mestizaje a la Diversidad Etnica y Cultural: la Contribucion del Movimiento Indigena y Negro de Honduras." In *Memorias del Mestizaje: Cultura Politica en Centroamerica de 1920 al Presente,* edited by Dario Euraque, Jeffrey Gould, and Charles Hale. Guatemala, Guatemala: Nawal Wuj S.A., 2005.

Brondo, Keri Vacanti. *Land Grab: Green Neoliberalism, Gender, and Garifuna Resistance.* Tucson: The University of Arizona Press, 2013.

Carey, David. "Hard Working, Orderly Little Women": Mayan Vendors and Marketplace Struggles in Early-Twentieth-Century Guatemala." *Ethnohistory: the Bulletin of the Ohio Valley Historic Indian Conference.* 55 (4) (2008): 579–607.

Central Intelligence Agency. "The World Factbook, Central America, Honduras." Last updated August 11, 2020. Accessed August 25, 2020. https://www.cia.gov/library/publications/the-world-factbook/geos/ho.html

Chenier, Jacqueline, Stephen Sherwood, and Tahnee Robertson. "Copan, Honduras: Collaboration for Identity, Equity, and Sustainability." In *Cultivating Peace: Conflict and Collaboration in Natural Resource Management,* edited by *Daniel* Buckles, 221–235. Washington, D.C.: International Development Center; World Bank Institute, 1999.

Euraque, Darío. "Antropólogos, arqueólogos, imperialismo y la mayanización de Honduras: 1890-1940." *Revista de Historia.* 45 (2002): 73–106.

Feldman, Lawrence. "Some Data and Reflections on the Dynamism and Ethnic Continuity of the Colonial Ch'orti' Population: The Many Copáns and San Juan Ermita." In *The Ch'orti' Maya Area, Past and Press,* edited by Brent Metz, Cameron McNeil and Kerry Hull, 148–156. Gainesville: University of Florida Press, 2009.

Ferguson, Lucy. 2010. "Interrogating 'Gender' in Development Policy and Practice." *International Feminist Journal of Politics* 12(1): 3–24.

García de Palacio, Diego, E. G. Squier, Alexander von Frantzius, and Frank E. Comparato. (1576). *Letter to the King of Spain: being a description of the ancient provinces of Guazacapan, Izalco, Cuscatlan, and Chiquimula, in the Audiencía of Guatemala, with an account of the languages, customs, and religion of their aboriginal inhabitants, and a description of the ruins of Copan.* Culver City, CA: Labyrinthos, 1985.

Gustafson, Bret. "Manipulating Cartographies: Plurinationalism, Autonomy, and Indigenous Resurgence in Bolivia." *Anthropological Quarterly* 82 (4) (2009): 985–1016.

Hale, Charles R. "Neoliberal Multiculturalism the Remaking of Cultural Rights and Racial Dominance in Central America." *The Political and Legal Anthropology Review* 28(1) (2005): 10–28.

Hawkins, Donald and Shaun Mann. "The World Bank's Role in Tourism Development." *Annals of Tourism Research* 34 (2) (2007): 348–363.

Herranz Herranz, Anastasio. "Los Mayas-Chorties de Honduras." *Mayab* 9 (1994): 87–92.

Hoksbergen, Roland and Noemi Espinoza Madrid. "The Evangelical Church and Its NGOs in Guatemala and Honduras." *The Journal of Developing Areas* 32(1) (1997): 37–52.

Junho, Maria-Valerica and Kathryn Johns Swartz. "Guiding the Way to New Opportunities in the Copan Valley, Honduras." *En Breve* 14 (2002): 1–4.

Little, Walter E. "Performing Tourism: Maya Women's Strategies." *Journal of Women in Culture and Society* 29 (2) (2003): 527–532.

Little, Walter E. *Mayas in the Marketplace: Tourism, Globalization, and Cultural Identity*. Austin, TX: University of Texas Press, 2004a.

Little, Walter E. "Outside of Social Movements: Dilemmas of Indigenous Handicraft Vendors in Guatemala." *American Ethnologist* 31 (1) (2004): 43–59.

López, Luis Enrique and Inge Sichra. "Intercultural Bilingual Education Among Indigenous Peoples in Latin America." In *Encyclopedia of Language and Education*, Vol 5, *Bilingual Education*, edited by J. Cummings and N.H. Hornberger, 295–309. Boston, MA: Springer, 2007.

Martinez Perdomo, Adalid. *La Fuerza de la Sangre Chorti*. San Pedro Sula, Honduras: Industrias IMET, S.A., 1997.

Medina, Laurie K. "Commoditizing Culture: Tourism and Maya Identity." *Annals of Tourism Research* 30 (2) (2003): 353–368.

Metz, Brent. "Questions of Indigeneity and the (Re)-Emergent Ch'orti' Maya of Honduras." *Journal of Latin American and Caribbean Anthropology* 15(2) (2010): 289–316.

Metz, Brent, Cameron L. McNeil, and Kerry M. Hull. *The Ch'orti' Maya Area: Past and Present*. Gainesville, FL: University Press of Florida, 2009.

Mortensen, Lena. "Copan Past and Present: Maya Archaeological Tourism and the Ch'orti' in Honduras." In *The Ch'orti' Maya Area: Past and Present*, edited by Brent Metz, Cameron L. McNeil, and Kerry M. Hull, 246–255. Gainesville, FL: University Press of Florida, 2009a.

Mortensen, Lena. Constructing Heritage at Copan, Honduras: An Ethnography of the Archaeology Industry." PhD diss., Indiana University, 2005.

Restall, Matthew, and Solari, Amara. *2012 and the End of the World: The Western Roots of the Maya Apocalypse*. Lanham, MD: Rowman & Littlefield Publishers, 2011.

Rivera, Mariel. *Fondo Prosperidad: Proyecto de Desarrollo Regional en el Valle de Copán*. Internal Report, *Secretaria de Turismo; Instituto Hondureño de Turismo*, Unpublished, 2013.

Rodriguez-Mejia, Fredy. "From Expeditions to Exhibitions: Circulation of Anthropological Knowledge in Copán, Honduras." *Anthropology News* 52(2) (2011): 10.

Rodriguez-Mejia, Fredy Articulations of identity through struggle among the Ch'orti' Maya of Copán, Honduras." PhD dissertation, Michigan State University, East Lansing, Michigan, 2016.

United Nations. "Inequalities in Human Development in the 21st Century: Briefing note for countries on the 2019 Human Development Report, Honduras," 1–10. Accessed August 25, 2020. http://hdr.undp.org/sites/all/themes/hdr_theme/country-notes/HND.pdf 2019.

Van den Berghe, Pierre. "Marketing Mayas: Ethnic Tourism Promotion in Mexico." *Annals of Tourism Research* 22(3) (1995): 568–588.

Yeager, Timothy J. "Encomienda or Slavery? The Spanish Crown's Choice of Labor Organization in Sixteenth-Century Spanish America." *The Journal of Economic History* 55(4) (1995): 842–859.

Yúdice, George. *The Expediency of Culture: Uses of Culture in the Global Era*. Durham, NC: Duke University Press, 2005.

# *Chapter 7*

# Aboriginal Alternative Tourism in Arnhem Land

## *Tourism as Cultural Learning*

Sachiko Kubota

Tourism in Australia has been seen as holding economic potential from the early 1900s, a relatively early stage and has been propelled enthusiastically. By the 1990s, it had become a more robust industry compared to the primary industries, which had been Australia's main export and income source for a long time. The Aboriginal content has remained a minor part of tourism for a long time (Altman 1989). However, it has coexisted with other attractions such as settlers' history, unique fauna, and landscape. The Australian National Travel Association (ANTA) was established in 1929 and functioned as a significant force in advancing Australian interests in the international tourist markets (Geissler 2019). For example, ANTA reported that the number of visitors from the United States was 27,675 in 1965,64,281 in 1970, and 83,000 in 1971 (Geissler 2019).[1]

In 1965, Australia's Travel and Tourist Industry Report drew attention to the Indigenous sector's importance and identified the great potential for cultural tourism (Craik 1997). The Australian Tourist Commission, which aims to increase international tourism, was established in 1967. Moreover, when the Labor government came into power in 1972, the Department of Tourism and Recreation was established, which enabled the government to pursue the national tourism strategy.

Altman (1988) noted that access to Aboriginal tourism in the Northern Territory was limited in the late 1980s. Since then, tourism has been promoted internationally, and Australia has become a popular holiday destination. Interest in the culture of Aboriginal people, who hold unique and traditional culture, has grown accordingly. They have gradually become the focus of Australian tourism, both domestically and internationally. "One of the world's oldest continuous cultures" is a common phrase to attract tourists' attention.

Cultural tourism is a well-studied topic in anthropology (Craik 1997; Chhabra, Healy & Sills 2003; Smith & Robinson 2006 among others). There has been some discussion about the potentiality of sharing of culture or experiencing culture as a study focus (Tonnaer 2010). Bruner (2005) developed the concept of the "touristic borderzone" (p.18). He argued that cultural creativity developed in the encounter between hosts, residents, and tourists and pointed out that tourism's cultural production should be considered an interactive process. He called such space the "touristic borderzone." Currently, such an encounter is growing; thus, the borderzone is emerging in every corner of the globe. Bruner (2005) defined the borderzone as "creative space, a site for the invention of culture on a massive scale" (Bruner 2005:193). Tonnaer (2010) developed the idea further and focused on the performative exchanges of cross-cultural creativity at the center of the tourist encounter, and called it "encounter culture," emphasizing the importance of the sharing culture as a focus of tourism studies (Tonnaer 2010:25–26).

The case I will focus on in this chapter is an example of sharing culture that provides a very dynamic effect. Due to the Aboriginal communities' remoteness, where relatively traditional life is kept, it is not easy for tourists to visit. Since the last decade or so, several unique tourism attempts have been in operation in the remote north, described as alternative tourism cases. This chapter will introduce one alternative tourism case and discuss its social contexts to examine its potentiality.

## SOCIAL BACKGROUND OF ABORIGINAL CULTURAL TOURISM

Aboriginal Australians went through harsh colonization experiences following the arrival of the first fleet from England in 1788. The population declined rapidly due to the colonizers' disease and violence. The estimated 500,000 in 1788 declined to 60,000 at the beginning of the twentieth century. Aboriginal Australians have been seen as *stone age people* for a long time, and their culture has not been valued. The public policy and Aboriginal people's treatment started to change in the middle of the twentieth century. Aboriginal peoples' miserable situation, especially in the south, started to be recognized as a social problem (Broome 2010).

Furthermore, at the beginning of the twentieth century, Christian missionaries expanded their activities to the north of the continent, where Aboriginal people lived a relatively traditional life in isolation. Townships were built to sedentarize, Christianize, and assimilate them. Schools were built to teach English, and various industries tried to make the townships self-sustaining (Harris 2013).

After World War II, the move for expansion of Aboriginal rights was accelerated, and in 1967, the national referendum changed Aboriginal people's status completely. The constitution recognized their achieved status as equal to other citizens. By the 1970s, policy on Aboriginal people had changed to "self-determination," and the missionaries in the north who had managed the Aboriginal communities left, handing over management to Aboriginal people (Broome 2010).

The *Yolngu* is one of the language groups whose area is in the northeastern Arnhem Land. Methodist missions set up the townships in the area and educated and assimilated *Yolngu* people. Since the 1970s, the *Yolngu* have managed their community autonomously. The alternative tourism case I describe in this chapter occurred in one of the communities in the *Yolngu* area.

Since the 1960s, Aboriginal protests and calls for justice and equality have grown. The government took various measures to equalize treatment toward Aboriginal people, including enacting the Aboriginal Land Right Act (Northern Territory [NT]) in 1976 (Broome 2010). The first legal court case—the Gove land rights case—happened in the *Yolngu* area in 1971 (Wells 1982). Their claim was dismissed, with the court noting that Australia was *terra nullius*, meaning that no one owned the land. But at the same time, the government established the Royal Commission to investigate the Aboriginal concept of land ownership. From the report, the Aboriginal Land Rights Act NT was proposed and passed in 1976.

The first Aboriginal legal land rights case attracted wide social attention, and along with the famous bark petitions, was sent by the *Yolngu* to the lower house. From that time, *Yolngu* became a focus as one of the main traditional Aboriginal people who was not mix blooded, with traditions, language, and culture. Also, with successful modern music bands and the images in feature films, the *Yolngu* are now well known as an Aboriginal group with tradition (Kubota 2020a).

Beside land rights claims, other Aboriginal issues were also in focus in the 1980s in Australia. On January 26, 1988, the 200th anniversary of the First Fleet's arrival in Australia, more than 40,000 Indigenous and non-Indigenous Australians from around the country gathered and marched through Sydney on what subsequently became known as the "Day of Mourning" demonstration. The march was the largest demonstration in Australian history. "White Australia has a black history" was the message, claiming fair treatment of Aboriginal rights. This incident made a deep impression on the Australian public and caused them to realize that they had to solve Aboriginal issues (Marcus 1988).

Furthermore, in the late 1980s, "Aboriginal Death in Custody" appeared as a social issue. It has long been known that the number of Aboriginal people in custody is disproportionately high. With the reports of a series of suspicious Aboriginal deaths in custody from 1980 to 1989, the Royal Commission was set up in 1989 to study ninety-nine cases. The report of the commission was submitted in 1990. Upon the recommendation, the Council for Aboriginal Reconciliation was established as a statutory body under the Council for Aboriginal Reconciliation Act of 1991. The council was to last for ten years, until January 1, 2001. The council served to promote social understanding of Aboriginal people's unique position and culture and to enlarge the commitment to address historical and current disadvantages of them as a process of reconciliation between Indigenous people and the broader Australian society (The Council for Aboriginal Reconciliation, n.d.).

The Stolen Generation was another issue concerning Aboriginal people that attracted wide attention in the 1980s. Aboriginal children, especially the mixed-blood children, were forcibly removed from their families and were put into institutions run by Christian missions or foster families for assimilation. Those people are called "Stolen Generations." The policy was realized under the acts of respective colonies started in the 1860s. This policy continued until the 1970s, and an estimated 100,000 children were affected.

In 1995, the National Inquiry into the Separation of Aboriginal and Torres Strait Islander Children from their families was held by the Human Rights and Equal Opportunity Commission. The Australian Human Rights Commission (n.d.) report, "Bringing them Home," was tabled in 1997 with several recommendations. One of the recommendations was a public apology from the federal government, which was denied by the Prime Minister John Howard. It evoked the national debate concerning apology. Accordingly, the reconciliation committee's focus was to include the apologies. Various events were planned to let the people show sympathy to Aboriginal people, especially to the survivors. The reconciliation committee tabled their final report on of December 27, 2000, a few days before the committee's closure on January 1, 2001. On the following day, the Corroboree 2000 event was planned. About 250,000 people from all over Australia gathered to cross the Sydney Harbor bridge expressing their feeling of sorrow and sympathy for Aboriginal people (National Museum Australia, 2020).

In 2008, after the general election held at the end of 2007, which resulted in John Howard losing the seat, a national apology was issued by the new prime minister, Kevin Rudd. He invited survivors of the Stolen Generations to the parliament, and big screens were set up in the parks and public areas all around Australia. A huge crowd gathered to listen to the historic apology. The speech was an honest, open-hearted message with sincere words of sorrow, which moved people emotionally and gave a strong impression. People hugged each other with tears, and a mood of forgiveness was shared. It was one of the most memorable incidents symbolizing reconciliation (Kubota 2020b).

After the national apology, the reconciliation committee ceased, and the Aboriginal healing foundation was set up to deal with Aboriginal people's historical trauma. Also, the nonprofit organization, Reconciliation Australia, was set up in 2001 as a successor to the reconciliation committee and continues various activities for realizing reconciliation (Reconciliation Australia, 2020).

## *MAWUL ROM* PROJECT

Arnhem Land is an area of about 10 square km located in the far north end of Australia. Under the Aboriginal Land Right Act (Northern Territory) in 1976, it was designated as the Aboriginal land trust, with a freehold title. The land is legally under perpetual lease, but the title is inalienable. In the Arnhem Land area, traditional Aboriginal culture, such as language, ritual, mythology, social organization, is kept relatively well. People who want to visit the area need to gain an entry permit from the community council. In addition, because it is very remote, people need to travel a distance in a light plane, which is quite costly in money and in time. In other words, there are apparent obstacles to tourism. However, a number of people have been eager to come to the area and several travel agencies supply such tours for the people on the consent of Aboriginal communities (Altman 1988). They are small group tours escorted by tour conductors, mainly targeting senior wealthy people. The following case is different from the above. Although there is some help from outside management to organize the tourism, it is planned and organized by Aboriginal people themselves, by the Aboriginal initiative.

Rev. Dr. Djiniyini Gondarra is quite a famous, high profile senior member of the *Yolngu* community. He is active in Church, taking the role of chairman of the Uniting Aboriginal and Islander Christian Congress and a community leader with strong leadership in the community. Also, for this chapter's context, it is essential to the point that he is a past member of the Australian Council for Aboriginal Reconciliation.

*Yolngu* reside in several settlements in northeast Arnhem Land. I have been continuing my relationship with the *Yolngu* family since 1986. Djiniyini has been discussing about the project to teach non-Aboriginal people about *Yolngu* knowledge for a long time. He has often said, "All these past years, we have been following *Balanda*[2] law. Now it is their turn to learn our law." Also, he has developed a plan for tourism to be introduced to generate income and make the community self-sustaining for a long time.

The *Mawul Rom* Project is a combination of Djiniyini's ideas of teaching law and tourism. The project was initiated among the family of Djiniyini, members of the *Dhurili* clan nation, and took shape with a *Balanda* coordinator from Darwin. In 2004, the pilot project commenced. About forty-three Indigenous and non-Indigenous attendants from all over Australia, participated, and local *Yolngu* people attended. It was held at Dhudopu homeland, on Elcho Island. The project is located about 200 km west of Gove, 400 km east of Darwin. The venue is a beautiful small homeland facing the sea with a large open area surrounded by bush land. It is owned by a man of the *Durili* nation whose wife is a niece to Djiniyini and to whom I was adopted. All the family members participated and collaborated.

Furthermore, the *Mawul Rom* Project officially started in 2006, and was held annually until 2012.[3] Each year, *Mawul Rom* had about 100 participants, Indigenous and non-Indigenous, from all over Australia. Some were from overseas. Activities were held with local *Yolngu*, especially the family members of Djiniyini (Cross-Cultural Mediation and Leadership Project 2007).

As Djiniyini explained, the purpose of the *Mawul Rom* Project was to create a cross-cultural learning space in Arnhem Land. Through the *Mawul Rom* Project, people were expected to be trained from two cultural perspectives in conflict resolution and leadership. Specifically, participants camp for seven days at Dhudopu outstation and learn about *Yolngu* law, especially about peacemaking and conflict resolution. The cost of attending the *Mawul Rom* was 2,500 A$ (US$1,883) for a week, including meals and camp accommodation.

In the project, local *Yolngu* people participated in the ritual and dance with the participants every day. The talks by elders delivered every evening on *Yolngu*'s world view, with perspectives on gender, kinship, mythology, and land. The project aimed to let the participants understand the procedure and the concept of a peacemaking ceremony (Newman 2004). *Mawul Rom* has been *Yolngu* tradition in their ritual for conflict resolution. He selected it for this project. Djiniyini explained about the *Mawul Rom* as follows (pers. com)

> When some dispute or conflict happens, we held *Ngarra* Ceremony. We decide how to deal with violence. *Wikindi Rom*, which is the part of *Ngarra* Ceremony, will be held. *Wukkindi* literally means "the blood." The blood is very sacred for *Yolngu*. When the conflict happens, and the sufferers' group is upset, their blood is boiling. We need to atone the blood of the suffered clan people. *Mawul Rom* will be held for this purpose. In *Wikindi Rom*, the people affirm that the people have an intimate relationship even if you are from different clans. Then in *Mawul Rom*, the arbitrators will mediate for conciliation.

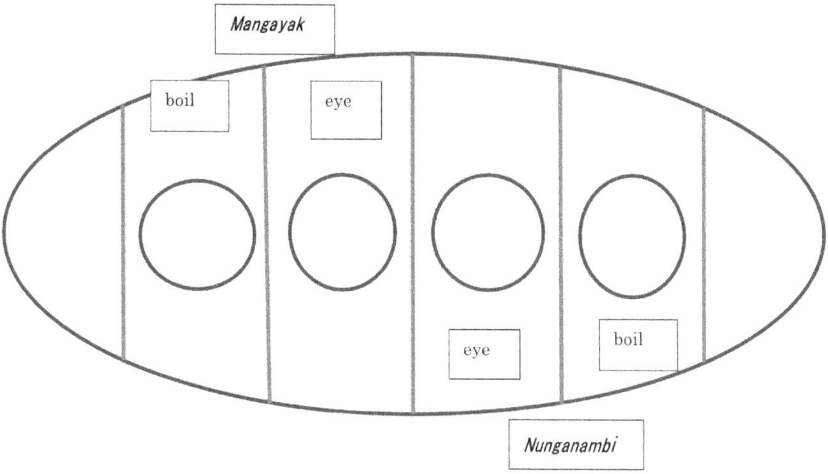


**Figure 7.1 Mok Design for *Mahwul Rom*.** *Source*: Sachiko Kubota.

According to Djinyini, some people are either *Nunganamb*i or *Mangayak* in the *Yolngu* world. They are like two different characters equally important as law-keepers, and symbolically they are called bloodlines. Each clan has both of them, although only a few, as their clan members. Only Law Man and Law Woman can identify *Nunganambi* and *Mangayak*. They have unusual eyes that can see through the person's reality. *Nunganambi* and *Mangayak* act as the mediators in *Mawul Rom*.

*Molk*, a sand sculpture with stringybark leaves, is built in the ceremonial ground center (see figure 7.1). The four circles made inside the *Molk* symbolize the eyes and boils of

*Nunganambi* and *Mangayak*. Left side circles represent a boil and an eye of *Mangayak*, and right side ones represent an eye and a boil of *Mangayak*. A boil of Mangayak at the far left is easy to press pus out of, but the other boil of Nunganambi at the far right is hard and very difficult to press out pus from. To make peace, people have to discard all the evil feelings, anger, and sorrow that are symbolized as rubbish, and they physically discard waste and rubbish into circles in the *Molk*. Two circles in the center represent the eyes of mediators, Nunganambi and Mangayak. Djiniyini explained how the ceremony proceeds (pers. com):

> People come into the ceremonial ground and sit around *Molk* quietly, sufferers' group on the one side and perpetrators' group on the other, together with mediators. They try to calm down and settle the anger without looking up to each other. Everyone look[s] down to the ground and trie[s] to reflect on themselves. Nobody speaks, and everyone sits quietly. The mediator will hand over meals to everyone,

> and people try to eat quietly without a word, still looking down self-reflexing. If anyone feels that he cannot control anger, [he] stand[s] up, and leave[s]. Mediators try very hard not to have anyone to stand up. The role of mediators is crucial. When everyone [finishes] meals, they . . . put the leftover food and bones [and] rubbish into *Molk*'s circles. Then the perpetrators' group will hand out water to the sufferers' group. If they accept and drink, then conciliation is settled.

As you can see, the concept of conflict resolution in *Mawul Rom* is unique. It is peace making by participation and emotional control. It requires peoples' deep understanding of *Yolngu* culture to be able to accomplish conflict resolution in this way (Briggs & Tonnaer 2008).

In 2011, *Mawul Rom* was formally designated as a program at Charles Darwin University. It is four years of education in law. Students are required to attend *Mawul Rom* annually for four years. By the end of four years, they will obtain a master's degree in *Mawul Rom*, with understanding of the Aboriginal way of making peace.

Especially with this change, many participants in the project are law-related personnel such as solicitors, law practitioners, law students. Nevertheless, many tourists who are not part of the university as well as workers in the public sector are attending too.

The *Mawul Rom* Project's final object is to let the participants understand the *Yolngu* way of peacemaking and resolution. For that purpose, the seven-day study is planned to teach them *Yolngu* life, philosophy, and world view. The narrative schedule of the seven days is shown in tables 7.1 and 7.2 based on my field research.

We can see that the participants are drawn to engage wholly in dancing, with paint all over their bodies to participate in the Rom ritual every day. At the same time, they are in intimate relationships with *Yolngu* people.

**Table 7.1 Abbreviated Schedule of the *Mawul Rom* Project**

| | *Day Time* | *Night* |
|---|---|---|
| Day 1 | Arrival, Registration | Prepare ground and *Molk* |
| Day 2 | First performance,<br>Dance | Body paint<br>Meal at *Molk*, self-reflection |
| Days 3 and 4 | Body paint, performance at *Molk* | Meal at *Molk*<br>Family adoption started |
| Days 5 and 6 | Body paint, performance at *Molk* | Meal at *Molk*<br>All the participants adopted into the families |
| Day 7 | Body paint (elaborate special one), Meeting by gender | Dance in the *Molk*,<br>Burn the culminated rubbish<br>Presents given<br>Departure |

*Source*: Author.

**Table 7.2 Narrative Schedule of the *Mawul Rom* Project**

| *Day* | *Description* |
|---|---|
| Day 1 | People arrive from all over Australia and are registered and introduced to the camping ground. Then, at night, senior members construct *Molk* at the center of the ceremonial ground. |
| Day 2 | It starts with a dancing session. The whole area is divided into two ceremonial grounds, including a place where *Yolngu* learning happens and the *Balanda* learning area under the tree. *Balanda* meditation and self-reflection are taught. In the afternoon, everyone is painted with red ocher and white clay and encouraged to dance. *Yolngu* culture is explained. In the evening, everyone takes a meal and sits down at the *Molk*. Djiniyini tells everyone to self-reflect and try to meditate, and experience with sense and feelings, not by intellect. Put the rubbish into the circle of the *Molk*. |
| Day 3 and 4 | In the morning, everyone is painted. They are taught the rules of the *Yolngu* world, including gender segregation. Participants are expected to learn to dance with *Yolngu* people. In the evening, senior members talk to everyone while taking meals and discarding rubbish into the circle. Participants start to be adopted into families. Skin names and marriage rules are explained. |
| Day 5 and 6 | In the morning, everyone is painted. Participants are expected to engage more and dance with the *Yolngu* for the ritual. Adoption is complete and explained. |
| Day 7 | In the morning, everyone is painted with elaborate designs. They are encouraged to dance around *Molk*, and in the end, all the rubbish is burned with branches. Participants are given presents (woomeras[a] for men, dilly bags[b] for women) from their family members, and then depart to home. |

[a] Woomeras are spear throwers made from trees. Participants use them for dancing too.
[b] Dilly bags.
*Source*: Author.

Participants learn about kinship and *Yolngu* world views, although sometimes challenging to understand. Participants emphasize the essential importance of immersing themselves in the *Yolngu* world. By doing so, it is expected that they get a unique experience that leads them to peacemaking in the *Yolngu* way (Briggs & Tonnaer 2008). They learn through direct contact with the *Yolngu* people sharing cross-cultural ideas, processes, symbols, and rituals related to peacemaking, and are expected to reach the understanding of conflict resolution in the *Yolngu* way in the final stage.

## CONCLUSION

In this chapter, I have focused on the unique case of alternative tourism in Arnhem Land, Australia. It was a project initiated by *Yolngu*, one of the language groups of Aboriginal people. From the examination, the expectation of

*Yolngu* toward the *Balanda* participants was made clear. They expect the participants to learn and acknowledge *Yolngu*'s knowledge and their way of life through whole bodily participation. The local *Yolngu* are willing to teach and share experiences concerning *Yolngu* life. For that purpose, it was imperative to apply the scheme of the *Yolngu* ritual of *Mawul Rom*.

Djiniyini's intention was clear. The planning of the project started in the early 2000s, just after the Corroboree 2000, and before the national apology. In other words, it was the time when the reconciliation with Aboriginal people was focused outstandingly in Australian social context. And Djiniyini had just finished his term as a member of the reconciliation committee. He knew there was a strong drive in the *Balanda* world for learning about the Aboriginal situation on site and also for seeking reconciliation. It was the reason he chose the *Mawul Rom*, a conflict resolution ritual, as the focus of this project. He took this opportunity to impose *Yolngu* knowledge, to let the *Balanda* people learn about peacemaking and conflict resolution in the *Yolngu* way.

As reconciliation was, and still is to some extent, one of society's main interests, many mainstream people from all over Australia participated in the event. In other words, the domestic circumstances surrounding Aboriginal people made this alternative tourism project possible. There are political and social needs in Australian society that supported it. The Aboriginal issue is crucially important in Australia, especially for many public sectors as well as private companies. Various governmental sections particularly need to deal with the Aboriginal issue with special care. Thus, there is a demand in mainstream society to understand Aboriginal people and their world view.

Moreover, a genuine appreciation of Aboriginal culture has been developed in the recent climate. Many mainstream Australians, especially those whose work deals with Aboriginal issues, sincerely appreciate Aboriginal culture. They are eager to learn about them and want to be sympathetic supporters. Moreover, by doing so, they heartily hope to solve the Aboriginal problems (Kowal 2015).

Djiniyini urged the participants to enroll in the project not by intellect but with full emotion and whole bodily participation. They were to penetrate the ritual by dancing together with *Yolngu* with paints on their bodies, physically surrounded by the nature of Arnhem Land, the bush, and the sea. In a way, the *Mawul Rom* Project provided the "cultural package" that enabled people to be part of the *Yolngu* way, including the sensual feeling and sentiment. By attending, eating, dancing, and sitting down on the ground with *Yolngu*, participants experienced whole bodily and emotional immersion.

As we have seen, the social circumstances in the 2000s surrounding Aboriginal people in Australia made this unique tourism project possible. For the participants who appreciated and valued the *Yolngu* cultural experience, it was a very rare and important opportunity. The *Mawul Rom* Project provided the cultural package experience where participants could penetrate emotionally and wholly bodily, and with dynamic interaction. With several social situations combined, participants could share culture in this dynamic interactive tourism space, where *Yolngu* and *Balanda*'s interaction was happening in the "border zone," the creative space in tourism as Bruner (2005:18) notes, but in the *Mawul Rom* Project, it was a more dynamic space that enabled the people to immerse themselves into the world. For that reason, it may have provided the space for real changes which lead people to mutual understandings and possible reconciliation.

## NOTES

1. The data of 1965 is from (Australian Tourist Commission 1971 Australian Overseas Visitor Statistics 1965–1970, Melbourne: Development & Research division, Australian Tourist Commission), and of 1971 is from (Australian Tourist Commission 1972 Australian Overseas Visitor Statistics 1966–1971, Melbourne: Development & Research division, Australian Tourist Commission), according to Geissler 2019.
2. *Balanda* is the *Yolngu* word which means non-Aboriginal people, mainly white Australians.
3. In 2012, the *Mawul Rom* Project was held at Mandorrah, Northern Territory. Because a young boy in a family died tragically rescuing a girl in the sea in front of Dhudupo homeland, the event's location was moved.

## REFERENCES

Altman, Jon. *Aborigines, Tourism, and Development: The Northern Territory Experience*. Australian National University North Australia Research Unit Monograph. Canberra: Australian National University, 1988.

Altman, Jon. "Tourism Dilemmas for Aboriginal Australians." *Annals of Tourism Research* 16 (1989): 436–476.

Altman, Jon, and Julie Finlayson. "Aborigines, Tourism and Sustainable Development." *Journal of Tourism Studies* 4(1) (1993): 38–50.

Australian Human Rights Commission. "Bringing them home report (1997)." https://humanrights.gov.au/our-work/bringing-them-home-report-1997.

Briggs, Morgan, and Anke Tonnaer. 2008. "Mawul Rom Project: Openness, Obligation and Reconciliation." *Australian Aboriginal Studies* 2 (2008): 3–15.

Broome, Richard. *Aboriginal Australians: A History Since 1788*. Sydney: Allen & Unwin Academic, 2010.

Bruner, Edward M. *Culture on Tour: Ethnographies of Travel.* Chicago: University of Chicago Press, 2005.

Chhabra, D., R. Healy, and E. Sills. 2003. "Staged Authenticity and Heritage Tourism." *Annals of Tourism Research* 30(3) (2003): 702–719.

The Council for Aboriginal Reconciliation (2014). http://www5.austlii.edu.au/au/orgs/car/council/index.htm

Cross-Cultural Mediation and Leadership Project. *Mawul Rom Project: Traditional and Contemporary Mediation and Leadership Training.* 2007. indigenousjustice.gov.au.

Geissler, Marie. "Cultural Tourism: Imagery of Arnhem Land Bark Paintings Informs Australian Messaging to the Post-War USA." *Arts* 8 (2019): 66.

Craik, Jennifer. "The Culture of Tourism." In *Touring Cultures: Transformations and Theory*, edited by Chris Rojek and John Urry, 113–136. London: Routledge, 1997.

Harris, John. *One Blood—200 years of Aboriginal Encounter with Christianity: A Story of Hope*. Bentford Square, Victoria: Australians Together, 2013.

Kowal, Emma. *Trapped in the Gap: Doing Good in Indigenous Australia.* Oxford and New York: Berghahn Books, 2015.

Kubota, Sachiko. "Seiji-teki Dainamizumu to Enkakuchi no Aborigini [Political Dynamism and Aboriginal People in Remote Area."] In *O-sutoraria Tabunka Shakairon* [*Australian Multicultural Society*], edited by Sekine, Masami et al., 79–93. "", Tokyo: Horitsu-Bunkasha, 2020a (in Japanese).

Kubota, Sachiko. Wakai toiu Michisuji no Kanousei [Possibility of the Reconciliation as a peace making], In *Buzokuhunnso Jokyo no Tagenteki Kenkyu [Reaserch from plural perspectives on the social situation after Ethnic Conflict*, edited by Tsukimori, T., 77–104. Kyoto: Koyo-shobo, 2020b (in Japanese). .

Marcus, Julie. "Bicentenary Follies: Australians in Search of Themselves." *Anthropology Today* 4(3) (1988): 4–6.

National Museum Australia. Walk for Reconciliation." 2020 https://www.nma.gov.au/defining-moments/resources/walk-for-reconciliation

Newman, Samantha. "The Mawul Rom Project: An Experience of Cross-Cultural Mediation." *Indigenous Law Bulletin* 6, no. 6 (2004): 11–13.

Reconciliation Australia. 2020. https://www.reconciliation.org.au/

Smith, Melanie K., and Mike Robinson. *Cultural Tourism in a Changing World — Politics, Participation and (Re)Presentation.* Clevedon: Channel View Publications, 2006.

Tonnaer, Anke. "A Ritual of Meeting: Sharing Culture as a Shared Culture in Australian Indigenous Tourism." *La Ricerea Folklorica* 61 (2010): 21–31.

Wells, Edgar. *Reward and Punishment in Arnhem Land, 1962-1963*. Canberra: Australian Institute of Aboriginal Studies, 1982.

*Chapter 8*

# How Does Cultural Tourism in a National Monument Affect Its Affiliated Pueblo Tribes?

Vidal Gonzales, Robert K. Hitchcock, and Melinda C. Kelly

Since 1872, the United States government has established protected areas, including national parks and national monuments, on the lands of Indigenous peoples (Keller and Turek 1999; Spence 1999; Burnham 2000). These protected area declarations have usually not come out of consultations with Indigenous peoples, who generally were not invited to participate in the planning and implementation of the protected areas. It was not until the twenty-first century that Indigenous communities were consulted formally about the establishment of parks and monuments, notably in the case of Bears Ears National Monument (BENM), which was declared in December 2016 by then-president Barack Obama. Bears Ears was the first native-proposed national monument in the United States (Wilkinson 2018).

In this chapter, we first discuss those Pueblo peoples in the Southwestern United States who are affiliated with Bandelier National Monument in north-central New Mexico. We examine these Indigenous peoples' complex relationships to Bandelier National Monument. Next, we define cultural tourism and its importance to Indigenous peoples' economic development. After that, we evaluate the effects of cultural tourism in Bandelier National Monument on affiliated Puebloan societies, economies, cultural resources, and cultural heritage. Attention is given to employment, job creation, and youth programs involving cultural heritage management. We close with a discussion of possible solutions to the problems of national monuments and protected areas vis-à-vis Indigenous peoples, including the issues of tribal co-management and specialized access to sacred sites and natural resources inside Bandelier.

## THE PUEBLOS OF THE SOUTHWESTERN UNITED STATES

Pueblo peoples are Indigenous peoples that reside currently in the American states of Arizona, Colorado, New Mexico, and Utah (Dozier 1983). They share common customs, traditions, religious, and other practices (Cordell 1979:131–133; Eggan 1979:224–254; Davis 1996:489–522). These Indigenous peoples have existed in the Southwest dating back 11,000 or more years (Powers 2005). The Ancestral Pueblo culture is responsible for creating the architectural marvels of Bandelier, Chaco Canyon, Mesa Verde, and Aztec, among others (Cordell 1979). In the 1200s and 1300s, Pueblo people started to settle the Rio Grande Valley in New Mexico, many of them creating the adobe brick and stone Pueblos and modern Puebloan cultural complexes that early Spanish expeditions encountered beginning in the 1540s (Dozier 1983:38–45, Baca 2015:49–55). Puebloan villages such as Acoma in western New Mexico and the Hopi Mesas in Arizona have remained almost permanently occupied since the eleventh century. Some of the Pueblos moved from the bottom of the mesas to the top in order to seek refuge from raids or to afford greater visibility (Dozier 1983:37).

Modern Pueblo culture is composed of highly complex social organizations and religious beliefs. Every Pueblo has different religious conceptualizations and social organization, but these institutions revolve around the same basic religious beliefs and cosmology (Parsons et al. 1996; Ortiz 2009). All Pueblo people have origin stories of ascending from an underground world and coming up into their homelands (Parsons et al. 1996: 210–218). Pueblo religion is based on a set of beliefs about close associations between people and nature, the importance of ancestors, communications with spirits, communal rituals, sacred places in residential areas (kivas) where ceremonies are held, and the importance of sacred sites and localities. Pueblo religion includes paying homage, making pilgrimages, praying to ancestral gods, and visiting places that their ancestors made sacred, which is significant in Pueblo religion today (Parsons et al. 1996). Kivas, above and below ground ceremonial chambers, are the center of much Pueblo religious life. Many Pueblo people hold beliefs that include enticing rain, interacting with animals, and ensuring good health and prosperity through dance, chants, prayer, and other religious acts. They believe their ancestors' pueblos to be places of utmost importance in correctly practicing their religion and rituals. Localities like Chaco Canyon, Mesa Verde, and Bandelier represent a category of elevated significance for Pueblo people. The same is true for sacred mountains such as Mount Taylor (called *Tzoodzil* by the Navajo, *Dwankwi Kyabachu Yalanne* by the Zuni, *Kaweshtima* by the Acoma; *Tsibina* by the Laguna; and *Tsiipiya* by the Hopi.), sacred lakes (e.g., Blue Lake, *Ba Whyea*, near Taos Pueblo), sacred springs (e.g., those in the Jemez Mountains), and sacred rivers (e.g., the Rio Grande River, *Posoge* in Keresan, *Paslápaane* in Tewa).

The Eastern Pueblos in the Rio Grande Valley have several unique features. Land tends to be owned by men, while women do much of the agriculture, food processing, and make pottery, and both men and women engage in house construction. Traditional governance was theocratic, in which priests and local leaders collaborated along with Pueblo community members. Much of the firewood and wild plant collection was done by women and hunting generally was done by men.

Decisions about moving to another location were usually done based on consensus, with adults taking part in discussions about various options. There were several reasons that might be given for a Pueblo being relocated: local soils were depleted judging from agricultural productivity, the area had become dry due to drought and climate change, or the location was seen as being subject to attacks by other groups. There is considerable debate in the literature about large-scale migrations that have occurred in the past (e.g., from the Mesa Verde area to the Rio Grande Valley) (Ortmann 2010, 2012, cf. Schillaci 2021). Some Pueblos, such as Jemez, moved short distances from their original locations to new ones only a few kilometers away for a combination of environmental, social, economic and political reasons (see, for example, Aiuvalasit 2017). Tourism has played a significant role among Native American communities, particularly those in the Southwest (see Sweet 1981, 1991, 2005; Laxson 1991). As will be shown below, the impacts of tourism on Pueblo societies and economies have been profound.

## THE ENVIRONMENT OF BANDELIER NATIONAL MONUMENT AND ITS AFFILIATED PUEBLOS

Bandelier National Monument (BNM) is named after the archaeologist and anthropologist Adolph Bandelier who worked in the area in the late nineteenth century (see Bandelier 1892). The Monument is located on the Pajarito Plateau, a large area comprised of mesas and canyons of volcanic tuff in north-central New Mexico. The Pajarito Plateau is a volcanic plateau whose elevations range from about 5,600 feet (1,700 meters) at the river to about 7,800 feet (2,377 meters) and is 385 $km^2$ in size. The plateau stretches from the Rio Grande River north and west into the Jemez Mountains, a prominent feature of which is the Valles Caldera, a volcanic feature that had major eruptions 1.6 and 1.2 million years ago that today is a National Preserve. Native peoples utilized the entire area and its natural resources, including wild animals, plants, and minerals such as volcanic glass, or obsidian (often used to produce projectile points and scrapers).

Today the average yearly amount of precipitation for the year in Bandelier National Monument is 15.4″ (391.2 mm). The month with the most precipitation is August with 2.8″ (71.1 mm) of precipitation. Rainfall varies spatially and over time, and droughts are not uncommon on the plateau (Bowen 1990; Allen 2004:22–28, 42–51). Vegetation includes grasslands, pinon-juniper woodlands, Ponderosa pine forests, and mountainous areas with white fir, aspen, and limber pine (Allen 2004:28–30). Fire is a crucial part of the ecology of the Pajarito Plateau, and was used by native peoples to manage plants, animals, and habitats. As Traynor et al. (1990) note, fire has both ecological and cultural resource impacts in Bandelier National Monument and surrounding areas. The animal population of the region is diverse, with some thirty-six species of mammals ranging from cottontail rabbits and squirrels to mule deer, elk, and black bears. The valleys in the plateau contain soils rich in nutrients which were and are utilized by native peoples for agricultural purposes.

Bandelier National Monument is a protected area under the management of the National Park Service (NPS) of the Department of the Interior (Rothman 1988; Barey 1990). The Monument incorporates part of the Pajarito Plateau, which is an area defined by dense Ancestral Puebloan dwelling sites and archaeological sites (Toll 1995; Stuart 1989, 2009, 2010; Kohler 2004a, b; Powers 2005). The Park's main attraction, Frijoles Canyon, contains many Ancestral Puebloan sites built by farmers who utilized the area over the past 1,500 years (Powers 2005:21–23, 36). Extensive surveys and excavations have helped to refine a chronology of occupation and answered key research questions about such issues as village formation, aggregation, and socioeconomic change (Kohler 2004a:2–6). This work, some of which was done in the 1990s, also served to shed light on key questions about the impacts of environmental change (Allen 2004). Bandelier National Monument has a long history as a subject of archaeological interest and research, which has aided contemporary Pueblo peoples in their efforts to learn more about their prehistoric and historic pasts (Mathien 2004).

In the early twentieth century, there was a move to create a national park to protect these sites and their resources from extractive industries (Smith 2002). Bandelier National Monument was created in 1916 by proclamation of President Woodrow Wilson to protect and preserve for public enjoyment and education the large Pueblo settlements and spectacular cave dwellings of the southern Pajarito Plateau (Toll 1995: vii). The Monument was created originally under the jurisdiction of the United States Forest Service (USFS), but later was transferred to the jurisdiction of the US National Park Service (NPS) in the Department of the Interior in 1932 (Rothman 1988). This transfer ended grazing rights within the park, as well as natural resource collection, hunting, and gathering activities of nearby Pueblo communities

(Smith 2002:15). These issues continue to concern the Pueblos in the Bandelier region.

The creation of a national monument essentially invalidated any land restitution attempts the Pueblos would have been able to make in the future regarding their Aboriginal lands on Bandelier National Monument because of its status as a protected area. Even though the creation of a national monument on Pueblo Aboriginal lands was not the original legal dispossession of this land by the US government, it made it nearly impossible for any future restitution of land within the park's boundaries to occur.

While the Pueblos did not legally own Bandelier National Monument, its cultural landscape "remained an important subsistence and sacred resource area for the Pueblo peoples" (Toll 1995:17). Today there are seven tribes that are closely affiliated with Bandelier National Monument: Keres-speaking Pueblos of Cochiti, San Felipe, Zia, and Kewa (Santo Domingo), and Tewa-speaking Pueblos of San Ildefonso, Santa Clara, and Ohkay Owingeh (San Juan) (Merlan and Levine 2000) (see table 8.1). These Pueblos remain culturally, religiously, and historically connected to Bandelier National Monument, and they seek to utilize land in the Monument for their ongoing traditions and customs. The Monument also included historic sites relating to the period of the Pueblo Revolt (1680–1692) when many sites were built on mesa-tops and in remote places (Silverberg 1970; Preucel 2002; Roberts 2004; Liebman and

**Table 8.1 Pueblos Affiliated with Bandelier National Monument, New Mexico**

| *Name of Pueblo* | *Language Spoken* | *Population Size (2022)* | *Comments* |
|---|---|---|---|
| Cochiti | Keres | 464 | On Rio Grande River south of Santa Fe |
| Kewa (Santo Domingo) | Tewa | 2,269 | East side of Rio Grande River, South of Cochiti, north of San Felipe |
| Okhay Owingeh (San Juan) | Tewa | 1,560 (6,748 in surrounding area) | East side of Rio Grande River, Close to Espanola |
| San Felipe | Keres | 2,584 | On Rio Grande River, North of Albuquerque |
| San Ildefonso (Powhoge) | Tewa | 803 | On Rio Grande River north of Bandelier |
| Santa Clara | Tewa | 2,021 | On the Rio Grande River, North of San Ildefonso |
| Zia | Keres | 939 | On the Jemez River northwest of Albuquerque |
| Total | 2 languages | 10,640 | Northern and Central New Mexico |

*Source*: By the authors from population data for the pueblos.

Preucel 2007; Wilcox 2009; Aguilar 2019; Aguilar and Preucel 2013, 2019). It is important to note that the area represents an ethnogeographical boundary that separates the Tewa aboriginal lands to the north from the Keres aboriginal lands to the south.

## CULTURAL TOURISM AND INDIGENOUS ECONOMIC DEVELOPMENT

Indigenous people have only relatively recently become the targets of tourism and ecotourism (Zeppel 2006; Fennell 2020). Cultural tourism, as defined by the International Cultural Tourism Charter, which was drafted by the International Council on Monuments and Sites (ICOMOS), states that it is

> essentially that form of tourism that focuses on the culture and cultural environments including landscapes of the destination, the values and lifestyles, heritage, visual and performing arts, industries, traditions and leisure pursuits of the local population or host community. Cultural tourism can include attendance at cultural events, visits to museums and heritage places and mixing with local people (Richards 2018). It should not be regarded as a definable niche within the broad range of tourism activities, but encompasses all experiences absorbed by the visitor to a place that is beyond their own living environment. (ICOMOS 2002:22)

In this way, cultural tourism is any tourism that seeks to indulge in the cultural activities, cultural landscapes, environment, and heritage of a local or host community that is different from the visitors' cultural environment.

The ICOMOS International Cultural Tourism Charter explores how countries should manage cultural tourism for the purpose of cultural awareness, cultural/heritage preservation, and for the enjoyment/education of visitors (ICOMOS 2002). The charter also posits that the management of cultural tourism should include Indigenous peoples at all levels of planning and implementation and that the benefits of cultural tourism should be distributed evenly to enhance a host community's socioeconomic status (ICOMOS 2002). The charter places a great deal of significance on how cultural tourism should be utilized to enhance the socioeconomic status of Indigenous peoples since poverty levels and socioeconomic status among Indigenous peoples are much lower than those of the general population of a nation (ICOMOS 2002, World Bank and International Monetary Fund 2011). In 2021, Indigenous people made up 4–6 percent of the world's population but comprised 10–12 percent of the world's most impoverished people (Mamo 2021). This poverty level has been caused by discrimination in employment and political participation, the lack of quality education among Indigenous communities, high levels of nutritional stress and morbidity, and the destruction of their traditional economies (Bodley 2015; Neely and Hume 2020). In attempting to aid the reduction of poverty among Indigenous people, this charter encourages nations and cultural tourism management institutions to create employment opportunities for Indigenous peoples, to hire Indigenous peoples as interpreters and presenters, and to include Indigenous peoples in the planning, implementation, and management of cultural tourism sites/programs.

Cultural tourism has been and continues to be a mechanism of economic development among Indigenous communities around the world. Cultural tourism among Pueblo Indigenous peoples has been a significant economic benefit since the early 1900s and a contributor to the annual income of many households. As Edward Dozier, from Santa Clara, states, "thousands of tourists visit the Pueblos; and they (Pueblo Indians) have discovered that handsome profits can be made from arts and crafts and that the recent development of recreational facilities on their reservations can be lucrative" (Dozier 1983:110). This statement shows that cultural tourism's economic power to create profit for Indigenous peoples has been recognized by these communities and can and should be implemented by Indigenous communities on their own terms. When economic endeavors are created solely by Indigenous communities themselves, they have the power to create prosperous economic ventures that protect their cultures, environments, and political organizations, and that help to increase their socio-economic and sociopolitical statuses (Dozier 1983:110; Gerberich 2007:75–86).

Gerberich (2007:75–86) argues that there are four factors that create sustainable tourism for Indigenous peoples. The first factor is that of cultural sustainability, which is developing economic tourism in a manner that educates the tourists of that local/Indigenous community's beliefs/culture, that does not negatively impact the culture of that community, and that allows local/Indigenous communities to control the speed of their economic growth and cultural evolution (Gerberich 2007:78). The second factor is environmental sustainability, which is developing tourism in an approach that seeks to limit the negative effects of tourism on an ecosystem (Gerberich 2007:78–79). The third factor is socio-economic sustainability; it is the application of tourism development in a manner that improves local/Indigenous communities' economic statuses and well-being, that equally distributes the profits of tourism, and that improves local/Indigenous communities' traditional economies, and does not destroy them (Gerberich 2007:79). The fourth and last factor is that of political sustainability, which is the collaboration of local/Indigenous communities in the implementation, management, and development of tourism by bureaucracies and other organizations (Gerberich 2007:79–80). These factors are important because if they are not followed properly, they can create tourism developments that are severely damaging to local/Indigenous communities' cultures, traditions, traditional economies, language, politics, and heritage. But when Indigenous peoples themselves develop their tourism economies they are, in many cases, successful because they are fully conscious about their cultures, environments, traditional economies, social welfare, and their political/traditional organizations (Dozier 1983:110; Gerberich 2007; Aguilar 2019).

Consultation with local/Indigenous communities is a major factor in the successful implementation of tourism development. In the United States, consultation is guided by various laws and regulations for the purpose of maintaining beneficial Government-to-Government relations (Barrow 2009:107) and is the basic means by which American Indian tribes and federal agencies approach and resolve differences in the application of policies and regulations. Government-to-Government consultation is recognition of tribal sovereignty and the unique legal and historic relationship between the US Government and Indian tribes shaped by treaties, congressional acts, court decisions, Executive Orders and other actions of the Executive Branch (Barrow 2009:107). This means that the federal government of the United States is required by law to enter into consultation with Indian tribes on certain federal actions, which are the projects, activities, and programs concerning the affiliated tribes. Some of the laws and regulations that guide US consultation are the National Historic Preservation Act of 1966, the National Environmental Policy Act of 1969, the American Indian Religious Freedom Act of 1978, the Archaeological Resource Protection Act of 1979, and the Native American Graves Protection and Repatriation Act of 1979 (Barrow 2009: 107, NPS 2018). In the United States, the federal government is legally bound to stay in compliance with the various laws and regulations that guide consultation. Nevertheless, the degree to which federal agencies follow the regulations regarding consultation with Indigenous communities varies considerably.

## THE EFFECTS OF CULTURAL TOURISM IN BANDELIER NATIONAL MONUMENT ON PUEBLOAN ECONOMIES, CULTURAL RESOURCES, AND HERITAGE

A heritage impact assessment (HIA) (see Kimball 2017) reveals that cultural tourism and ecotourism in Bandelier have had both positive and negative effects on affiliated Pueblo economies, cultural resources, and heritage. The number of tourists who visited Bandelier National Monument between 2017 and 2021 is as follows (data from Bandelier National Monument and the New Mexico Tourism Department):

2017 — 209,141
2018 — 198,441
2019 — 200,741
2020 — 94,948
2021 — 270,716

According to a report from the National Park Service in 2021 tourism brought in $24,800,000 in economic benefits and provided 254 jobs in the local economy. A total of $18,963,000 was spent in nearby communities.

In understanding economic development, one can start to evaluate the positive and negative impacts of cultural tourism in Bandelier National Monument on its affiliated Puebloan economies. Bandelier National Monument has contributed positively to the increased economic development of Pueblos, both indirectly and directly. The status of Bandelier National Monument as a national monument brings tourists from all around the world to Pueblo country. Along with Mesa Verde National Park, it is one of the most heavily visited of the protected areas in the southwestern United States. The economic impacts of cultural tourism attractions on local communities stem from the associated expenses that tourists have while visiting, such as lodging, food, recreation, and shopping, which lead to profit among local businesses as well as increased employment levels (Boyd and Timothy 2006). In 2017, New Mexico had 2,045,049 National Park visits, with Bandelier National Monument being one of the top four protected areas to be visited (New Mexico Tourism Department 2018: 27).

Bandelier National Monument allows Pueblo artisans/craftsmen to sell their products every weekend at the park's visitor center and schedules weekend cultural demonstrations with Puebloan artisans, craftspeople, and dancers between Memorial Day and Labor Day (U.S. National Park Service *Cultural Demonstration Program,* 2020). The cultural demonstrators are paid by Bandelier National Monument for their time and work at the park, which also provide tourists with an understanding of genuine Pueblo culture and heritage. Bandelier National Monument also directly employs Pueblo people through the Bandelier Preservation Corps. (BPC) and through seasonal/permanent National Park positions, which are available because Bandelier National Monument is a cultural tourist destination. These types of indirect and direct economic impacts have helped to produce income among the affiliated Pueblos.

Bandelier National Monument has also negatively impacted the economic development of the Pueblos. Its designation as a protected area places restrictions on the use and collection of the cultural and natural resources of the park (Lott 2020). The resources once used by the Pueblo people to create pottery, buckskin clothing, moccasins, baskets, and other cultural items can no longer be exploited within the park's boundaries (Lott 2020), which has forced the Puebloan people to find those resources elsewhere. Bandelier National Monument does offer employment for Pueblo people, but it is very limited. In the summer of 2020, for example, Bandelier National Monument only had one permanent and two seasonal Puebloan employees. In 2022, it had five seasonal Puebloan employees (data from Bandelier National Monument, October 2022).

Through the BPC program Bandelier National Monument helps to fund the wages of approximately ten Puebloan youths from the various affiliated Pueblos for two-and-a-half-month seasonal jobs (Jadrnak 2015), but the Monument is unable to directly employ a sizable number of Pueblo people, which could have a considerable effect on Puebloan economies.

The cultural heritage resources of Puebloan Indigenous communities have been both positively and negatively affected by cultural tourism in Bandelier National Monument. Cultural resources can be broadly defined as all the tangible and intangible elements of a people's heritage (ICOMOS 2002:22). In a more restricted understanding, it can refer to the "physical evidence or place of past human activity: site, object, landscape, structure; or a site, structure, landscape, object or natural feature of significance to a group of people traditionally associated with it" (NPS 2015: *Cultural Resources*). This restrictive definition refers to cultural resources as only the tangible elements of a people's heritage and disregards the intangible elements of a people's heritage. For the purposes of this chapter, it is the restrictive definition which this chapter will utilize to understand cultural resources.

The National Park Service protects and preserves the cultural resources of the Pueblos within its boundaries through various laws and regulations that work to maintain their cultural and scientific significance (NPS 2018). Laws that guide these endeavors include the Antiquities Act of 1906, the Archaeological and Historic Preservation Act of 1960, the National Historic Preservation Act of 1966, and the Native American Graves Protection and Repatriation Act of 1990. Together, they serve to protect cultural resources from destruction, vandalism, theft, and other destructive acts (NPS 2018). The Bandelier Preservation Corps also works to physically preserve the Puebloan cultural resources of the monument. The BPC and its preservation department work on the pre-historic Ancestral Puebloan dwelling sites to repair and preserve their naturally and anthropogenically caused deterioration (Friends of Bandelier 2020).

All these protective/preservationist policies were made possible by the passing of the Organic Act, which allowed the creation of the National Park Service in 1916 (NPS 2018:9). The mission statement of NPS is,

> The National Park Service preserves unimpaired the natural and cultural resources and values of the national park system for the enjoyment, education,

> and inspiration of this and future generations. The Park Service cooperates with partners to extend the benefits of natural and cultural resource conservation and outdoor recreation throughout this country and the world. (NPS 2016)

Bandelier National Monument is a protected area that works to preserve the Puebloan cultural resources for the purpose of encouraging cultural tourism so that visitors may enjoy their visit and be educated and inspired by the park's cultural resources and the Pueblo people who work there as artists, musicians, and dancers.

Although the mission of NPS is geared to protect the natural and cultural resources of the United States, this same mission can lead to the destruction of those very resources by the promotion of enjoyment, education, and inspiration through the mechanism of cultural tourism due to the physical impacts that visitation has on these resources. In Bandelier National Monument, some of the negative physical impacts of visitation/cultural tourism on the cultural resources of the park include vandalism, heavy daily visitor traffic, inadvertent destruction of valuable sites, and theft. Vandalism at the monument in the form of purposeful graffiti or unintentional harm to shrines occurs almost daily. The constant traffic of visitors and visitor exploration causes the fragile cavates (homes excavated in volcanic cliffs) and former sites of the Ancestral Puebloans, which were both created out of soft volcanic tuff, to erode and become very unstable. Another problem in the park is the theft of Pueblo cultural resources, including pottery sherds, projectile points, and grindstones. These physical impacts by tourists have led directly to the destruction of Puebloan cultural resources.

The cultural resources of Pueblo people are being protected and preserved through NPS's management of Bandelier National Monument. However, this same management plan has resulted in the physical destruction and deterioration of these cultural resources. NPS must work to educate and inspire the public through the mechanism of cultural tourism by consulting with its affiliated Pueblo tribes to build a better management plan that balances their dual mission of protecting/preserving and that of educating/inspiring, so that the cultural resources of Pueblo people may be preserved effectively.

Local Pueblo members agree that the heritage of Pueblo Indigenous communities has been both positively and negatively affected by cultural tourism in Bandelier National Monument, depending in part on how one defines heritage. Heritage is a term that many people employ in interpreting and understanding a diverse range of concepts, from tangible materials to intangible practices that contribute to a relationship to the past (Harrison 2013:14). Heritage differs in connotation and meaning from community to community and from culture to culture, which complicates the issue of creating a homogenous sense of heritage and creates spaciotemporal conflicts between these various understandings of heritage (Harrison 2013:13–15). To be as inclusive as possible, heritage must be understood as the all the intangible and tangible practices, customs, places, objects, arts, and values that establishes a people to their past, ancestors, histories, and historical landscape (Harrison 2013:14; ICOMOS 2002:21–22).

Cultural tourism in Bandelier National Monument has had a positive effect on the heritage of Pueblo people by educating the broader public about Ancestral Puebloan culture and their descendants through the park's guided tours, websites, pamphlets, publications, and cultural demonstrations (NPS 2020). Heritage education brings awareness, appreciation, and understanding to the visitors of a heritage site that can facilitate respect and encourage good perceptions about the local or Indigenous populations that are affiliated with that heritage site (ICOMOS 2002; Goral 2011).

Another positive impact that Bandelier National Monument has had on the heritage of Pueblo peoples is the upholding of consultation laws, such as the American Indian Religious Freedom Act of 1978, which gives American Indians the right to access their religious, sacred, and ceremonial sites on federal lands (Barrow 2009; NPS 2018). This is important for Pueblo people because there are many shrines, Ancestral Pueblos, and sacred sites on Bandelier National Monument that continue to play a vital role in the ongoing traditions, practices, and heritage of today's Pueblo people. These sacred places must be visited, protected, refurbished, and venerated by Pueblo people for their heritage and cosmological understandings to continue. According to some members of local Pueblos, the Forest Service and later the National Park Service sought to stop visits to sacred sites in the protected area. In the 1930s, the last Pueblo households were removed to places outside of the monument's boundaries. This can be compared to the Chaco Culture National Historical Park, which removed the last Navajo family from the area in 1947 (Two Bears 2019).

The final positive impact on Pueblo people's heritage is the creation of the Bandelier Preservation Corps, which is composed of Puebloan youth and mentored by Pueblo adults who work for NPS (Jadrnak 2015). The BPC not only works to preserve their cultural resources (tangible heritage) but also serves to educate the public and the Pueblos about their heritage and ancestors who lived in Bandelier National Monument. This program has allowed for the traditional flow of traditional knowledge from elders to the younger generations, all while working to preserve their tangible heritage.

However, Pueblo people are no longer able to visit, pray, and make traditional pilgrimages without the permission of NPS to access sacred areas that are closed to public access in order to regulate the negative physical impacts of cultural tourism. Many Pueblo people do not understand the administrative process of obtaining permission to access these closed-off areas, which prevents them from conducting these practices and directly affects the traditional flow of knowledge from elders to youth, leading to the loss of heritage.

Cultural tourism in Bandelier National Monument has given NPS the tools to protect and preserve the heritage of Pueblo people through educating the public, upholding consultation laws/regulations, and by creating the BPC program. However, the tangible heritage of Pueblo people is deteriorating, being defaced, and being stolen by tourists who visit Bandelier, which also causes a disruption in the intangible heritage of the Pueblos Peoples. This disrupts their traditional flow of heritage from older generations to younger generations.

## THE BANDELIER PRESERVATION CORPS

The Bandelier Preservation Corps has been mentioned above, but it must be addressed because of its positive impacts on Pueblo peoples' economics, cultural resources, and heritage. The BPC is a flagship program that has brought together Indigenous mentors and Indigenous youth to accomplish the job of historic preservation (Jadrnak 2015). The BPC works to preserve the Ancestral Puebloan dwellings and other sites of Bandelier National Monument, some of which were constructed by their Tewa and Keres-speaking Puebloan ancestors (Merlan and Levine 2000). The BPC is a partnership between the National Park Service and the Rocky Mountain Youth Corps (RMYC), which works to create a youth work force that consists of Pueblo youth who are the descendants of the people that built the structures in Bandelier National Monument (*Bandelier National Monument: National Trust for Historic Preservation* 2015). The BPC program offers scholarships to Puebloan youth employees who can complete the two-and-a-half-month season, which has encouraged many BPC members to further their education. This means that Bandelier National Monument has been able to positively impact the socioeconomic status of Pueblo youth by offering them an income and a scholarship to an accredited institution. Another partnership with the American Conservation Experience (ACE) provides a full scholarship to a native Pueblo youth to work at Bandelier National Monument.

The BPC has provided a platform for Pueblo people to interpret their own cultural heritage to the public, which is not often done at national parks in the United States. The BPC works on the sites that are frequently visited by the public and this has led to a public-Pueblo interpretation mechanism. Allowing affiliated Puebloan people to interpret their own heritage to the public can result in visitors becoming more understanding and empathetic toward Indigenous peoples and their efforts to preserve their heritage

(ICOMOS 2002; Goral 2011). Furthermore, the employment of Pueblo people onto the BPC and NPS workforce has created a powerful consultation mechanism as NPS employees are exposed to Pueblo cosmological interpretations/beliefs and as Pueblo employees act an as internal checks-and-balances for NPS plans and policies. It would be useful for programs like the BPC to be implemented in national parks and other land management bureaucracies across the United States as they offers advantages to both the agency and to Indigenous communities and serve to create a viable and diverse workforce that can tackle the many laborious jobs that a land management agency needs to accomplish.

## POSSIBLE RESOLUTIONS TO THE NEGATIVE IMPACTS OF CULTURAL TOURISM IN BANDELIER NATIONAL MONUMENT

Internal consultation and tribal co-management are two possible resolutions to the negative impacts that cultural tourism in Bandelier National Monument brings to Pueblo economies, cultural resources, and heritage. Both resolutions are compatible and possible through the consultation laws/regulations that are afforded to tribes (Barrow 2009). In many cases, land management agencies do the bare minimum to comply with consultation laws (King 2013). This creates problems with the agencies and the affiliated tribes that occupy the area. These resolutions work to fix these issues, as well as the issues that cultural tourism in Bandelier National Monument brings to Pueblo economies, cultural resources, and heritage.

Internal consultation can be useful in making decisions about cultural issues in national parks and monuments, and can effectively raise the socioeconomic status of Bandelier National Monument's affiliated Pueblo tribes by offering tribal members federal jobs. Federal jobs can provide employees with stable incomes, quality health insurance, training, pensions, and other benefits, including networking opportunities, which can further their careers and professional development.

Hiring affiliated Pueblo tribal members onto the preservation management team of the Monument can help improve the physical preservation of Pueblo cultural resources. Pueblo tribal members bring traditional architecture and construction knowledge to the preservation team, which allows people to assess structural damage and fix it as their ancestors would have done to preserve the authenticity and quality of the site. Hiring affiliated tribal members also means that tribal members can provide internal checks-and-balances within the agency, while planning, designing, implementing, and managing the cultural resources of the Pueblos. Internal consultation effectively upholds the interests of the tribes by providing countless opportunities to add, alter, and change cultural resource management plans (Barrow 2009). This can serve to build an NPS workforce that is an ally to the preservation of Pueblo heritage and allows for the transfer of heritage from older Puebloan NPS employees to younger Puebloan NPS employees. Ultimately, internal consultation aids in the continuation of Puebloan heritage and is a tool that should uphold Pueblo Indigenous rights.

Tribal co-management means a "genuine sharing of power between equal partners, a nation-to-nation relationship based on co-existence," in which the government and a sovereign tribe co-operatively participate in the stewardship of a landscape (Rodon 1998:121–124). Based on this understanding, tribal co-management is the equal participation of the planning and implementation of management schemes by both the governmental agency and the tribal sovereign nations (Rodon 1998; Washburn 2021). Essentially, tribal co-management should not resemble an asymmetrical relationship where the governmental agency has the full authority over a landscape but should instead resemble a relationship in which both parties are equally as powerful and responsible (Rodon 1998). This will require a restructuring of land management agencies so that the decision-making process of a landscape will incorporate both entities on equal footing.

Tribal co-management has been occurring in Canyon de Chelly National Monument between the NPS and the Navajo Nation, which serves as an example that tribal co-management is possible in the United States (King 2007). The responsibilities of the park have been divided between the two entities. Essentially, the NPS manages the cultural and historic resources, while the Navajo Nation manages the natural resources of the monument, has accommodations for visitors, and handles transportation services (King 2007). Tribal co-management can also be achieved through the creation of an affiliated tribal management team that works side by side with an agency to implement, plan, and construct management schemes for an area. The affiliated tribal management team will be the directors of their own tribal staff, whose organization and structure will mirror that of the governmental agency they are working with to ensure that tribal consultation and co-management are occurring at all levels of the agency. Table 8.2 presents data on protected areas in the Southwestern United States in which co-management arrangements have been initiated.

The Bureau of Land Management (BLM) and Cochiti Pueblo have signed cooperative agreements for the management of the Kasha-Katuwe Tent Rocks National Monument in north-central New Mexico. Under these agreements, the Cochiti Pueblo people have significant management responsibilities for visitor liaison work, trail maintenance, law enforcement coordination with the BLM, and full-time staff recruitment to manage and monitor the

**Table 8.2 National Monuments and Historical Parks in the Southwestern United States**

| *Name and State* | *Date of Establishment* | *Size (km²)* | *Number of Affiliated Tribes* |
|---|---|---|---|
| Bandelier National Monument, New Mexico | 1916 | 136.29 km² | 7 |
| Canyon de Chelly National Monument, Arizona | 1931 | 339.5 km² | 2 |
| Bears Ears National Monument, Utah | 2016 | 1,351,849 acres, 5,470.74 km² | 5 |
| Chaco Culture National Historical Park, New Mexico | 1907 (redesignated as a National Historical Park (1980) | 137.5 km² | 25 |
| Mesa Verde National Park, Colorado | 1906 (29 June) | 52,485 acres (21,240 ha), 212.4 km² | 26 |
| Total | 5 | 6,296.43 km² | |

*Note*: Data obtained from the National Park Service, US Department of the Interior, Washington, DC, and from the websites of the individual monuments and national parks and specific tribes.

monument with BLM funds. Cochiti Pueblo is also part of the Inter-Tribal Bison Cooperative (ITBC) and manages its own buffalo herd.

Many of the Pueblos as well as the Diné have agreements with the US Fish & Wildlife Service regarding the handling of permits for fishing and hunting on Indian land. They collaborate on such matters as setting quotas and engaging in law enforcement (Richardson 2016; Leshy 2022a, b). They also interact with the US Fish and Wildlife Service on issues surrounding endangered species and the Endangered Species Act (ESA) on native land (Zellmer 1998).

Tribal co-management can effectively raise the socioeconomic status of Bandelier National Monument's affiliated Pueblo tribes by offering their tribal members jobs, access to vending, access to resources, and, through collaborative agreements, enabling Pueblo members access to sacred sites within the Monument. Job creation for Pueblo people at Bandelier National Monument has been accomplished through the creation of the tribal management team and, its employees, which works with NPS to manage the park.

In order to create better opportunities for Pueblo members to display and sell their work at the monument, the tribal management team could work to create a permanent space or facility for Pueblo artisans/craftspeople to sell their work directly to the tourists. Pueblo people have understood the economic viability of Pueblo arts and crafts since the early 1900s (Dozier

1983). Many Pueblos create cultural centers/events and allow vending at their traditional Pueblo feast days for artisans/craftsmen to sell their work to tourists (Dozier 1983). The Pueblo management team could also work toward loosening the restrictions on resource collection for Puebloan use of natural resources needed to create to create pottery, buckskin clothing, moccasins, baskets, and other cultural items.

Indigenous people often push for stricter standards of cultural resource preservation because there is "deep-seated interdependence between the intangible cultural heritage and the tangible cultural and natural heritage" of Indigenous peoples (UNESCO 2003:3). These stricter policies, regulations, plans, and implementation methods will result in lengthened survival of resources over the long term. Higher standards for the preservation of Puebloan cultural resources can aid in decreasing the negative physical impacts of cultural tourism.

Internal consultation and tribal co-management are consultation methods that actively involve the affiliated tribes of a land base in its care-taking. In the United States, tribal consultation is guided by the various laws that require the federal government to consult with tribes on certain federal actions that concern them (Barrow 2009). Many federal agencies currently allow tribes minimal input and power over the management of their affiliated objects, sites, and landscapes, but they could provide tribes with more power, authority, and capability to administer their own economies, cultural resources, and heritage (King 2013).

## CONCLUSIONS

In evaluating the impacts of cultural tourism in Bandelier National Monument on the economies, cultural resources, and heritage of this protected area's affiliated Pueblo tribes, we have attempted to show that there are ways to improve the relationships between the protected area and affiliated tribes and to mitigate negative tourism impacts. Pueblos and other Native Americans have a love-hate relationship with tourism. They value the opportunity to earn income, while they are concerned about the presence of tourists, especially those who arrive in large numbers. On balance, it appears that tourism currently imposes more costs than benefits on the Pueblos. We examined these Indigenous peoples' complex relationship with Bandelier National Monument and showed the various ways in which tourism could be improved. Next, we defined cultural tourism and its importance to Indigenous peoples' economic development, cultural resources, and heritage preservation. Co-management between the Pueblos and the National Monument is presented as a viable means of enhancing the viability of both the Pueblos and the protected area. Proposals regarding strategies for dealing with large numbers of tourists and their effects on cultural heritage have been suggested (Frey and Briviba 2021). One suggestion is to have small numbers of tourists participate in astronomical tourism or "dark sky tourism" in Bandelier, which also occurs in Chaco Canyon National Historical Park.

Bandelier National Monument should not be seen as an "island" but rather as part of a cultural landscape that is associated with Pueblo peoples. Some of the localities in and around the park demonstrate the effects of cultural and environmental change, including the effects of various systems of fire management, some governmental and some Indigenous. Lessons can be learned about how best to handle fires and ensure that local trees and other vegetation are not destroyed but rather protected and managed over the long term.

The Bandelier Preservation Corps is a highly useful flagship program that enhances the socioeconomic status of Puebloan youth as they work to preserve Puebloan cultural resources and heritage at Bandelier. Internal consultation and tribal co-management are resolutions that follow US consultation laws and procedures and are up to international standards. Enhanced consultation, participation, and co-management would go a long way toward resolving some of the negative impacts that cultural tourism in Bandelier National Monument brings to Pueblo economies, cultural resources, and heritage. Fundamentally, we call for cultural heritage and land-based management agencies to allow Indigenous peoples more control and power over their own heritage and natural and cultural landscapes. Facilitating more Indigenous people in their efforts to manage their own heritage and economic systems has the potential to create sustainable tourism, which is vital for the conservation of the environment, Indigenous economies, and Indigenous heritage.

## ACKNOWLEDGMENTS

An earlier version of this chapter was presented at the 78th Annual Plains Anthropological Conference, Boulder, Colorado, November 10–14, 2021. We would like to thank the tribal councils and members of the Pueblos of Cochiti, San Felipe, Kewa, Ohkay Owingeh (San Juan), Santa Clara, San Ildefonso, and Zia Pueblos who provided us with insights into their associations with Bandelier National Monument. We also thank the Southwest Regional Office of the National Park Service in Santa Fe. Support of some of the research for this chapter was provided to Vidal Gonzales through the University of New Mexico, Ohkay Owingeh, and the Santa Clara Pueblo Development Corporation. We thank Joseph P. Aguino, Governor of Ohkay Owingeh, Ann Braswell, Belle Cost, Christopher Chavez, Klara Kelley, Tiffany Lee, Joan Mathien, Alan Osborn, Alison Rautman, Matthew Schmader,

David Stuart, Richard Taylor, Chip Wills, and the late Alfonso Ortiz and the late Marta Weigel for their useful insights and information.

## REFERENCES

Aguilar, Joseph R. (2019). Asserting Sovereignty: An Indigenous Archaeology of the Pueblo Revolt Period at Tunyo, San Ildefonso Pueblo, New Mexico. PhD Dissertation, University of Pennsylvania, Philadelphia.

Aguilar, Joseph R., and Robert W. Preucel. (2013). Sacred Mesas: Pueblo Time, Space, and History in the Aftermath of the Pueblo Revolt of 1680. In *The Death of Prehistory*, edited by Peter Schmidt and Stephen A. Mrozowski, pp. 267–289. New York: Oxford University Press.

Aguilar, Joseph R. and Robert W. Preucel. (2019). Seeking Strength and Protection: Tewa Mobility during the Pueblo Revolt Period. In *The Continuous Path: Pueblo Movement and the Archaeology of Becoming,* edited by Samuel Duwe and Robert W. Preucel, pp. 149–165. Tucson: University of Arizona Press and New York: Tucson. American Museum of Natural History.

Aiuvalasit, Michael. (2017). Common Goods in Uncommon Times: Water, Droughts, and the Sustainability of Ancestral Pueblo Communities in the Jemez Mountains, New Mexico, AD 1100-1700. PhD Dissertation, Southern Methodist University, Dallas, Texas.

Allen, Craig D. (2004). Ecological Patterns and Environmental Change in the Bandelier Landscape. In *Archaeology of Bandelier National Monument: Village Formation on the Pajarito Plateau,* New Mexico, edited by Timothy A. Kohler, pp. 19–68. Albuquerque: University of New Mexico Press.

Baca, Jacobo D. (2015). Somos Indígena: Ethnic Politics and Land Tenure in New Mexico 1694-1965. Ph.D. Dissertation, Department of History, University of New Mexico, Albuquerque, New Mexico.

Bandelier, Adolph E. (1892). *Final Report of Investigations among the Indians of the Southwestern United States, Carried on Mainly in the Years from 1880 to 1885.* Papers of the Archaeological Institute of America, American Series, No. 4. Cambridge: Archaeological Institute of America.

Barey, Pamela. (1990). *Bandelier National Monument.* Tucson: Western National Parks.

Barrow, John M. (2009). *Preservation and Management Guidelines for Vanishing Treasures*, Intermountain Region of the National Park Service. Professional Paper No. 75. Intermountain Cultural Resource Management. National Park Service, U.S. Department of the Interior.

Bodley, John H. (2015). *Victims of Progress,* 6th edition. Rowman & Littlefield.

Bowen, Brent M. (1990). *Los Alamos Climatology.* Los Alamos National Laboratory LA-11735-MS-UC902. Los Alamos New Mexico: Los Alamos National Laboratory.

Burnham, Philip. (2000). *Indian Country, God's Country: Native Americans and the National Parks.* Washington, D.C. and Covelo, CA: Island Press.

Cordell, Linda S. (1979). Southwest. In *Handbook of North American Indians, Volume 9,* edited by William C. Sturtevant, pp. 131–151. Washington D.C.: Smithsonian Institution.

Davis, Mary B., ed. (1996). *Native America in the Twentieth Century: An Encyclopedia*. New York and London: Garland Publishing.

Dozier, Edward P. (1983). *The Pueblo Indians of North America*. Long Grove, IL: Waveland Press.

Eggan, Fred. (1970). Southwest. In *Handbook of North American Indians, Volume 9,* edited by William C. Sturtevant, pp. 224–235. Washington, D.C.: Smithsonian Institution.

Fennell, David. (2020). *Ecotourism*, 5th edition. London and New York: Routledge.

Franz, Daniel. (2021). The Subdelegation Doctrine as a Legal Tool for Establishing Tribal Comanagement of Public Lands: Through the Lens of Bears Ears National Monument. *Colorado Natural Resources, Energy, and Environmental Law Review* 32(1): 2–40.

Frey, Bruno S. and Andre Briviba. (2021). A policy proposal to deal with excessive cultural tourism. *European Planning Studies* 29(4): 601–618.

Gerberich, Victoria L. (2007). An Evaluation of Sustainable American Indian Tourism. In *Indigenous Tourism: The Commodification and Management of Culture*, edited by Chris Ryan and Michelle Aicken, pp. 75–86. London: Routledge.

Goral, Anna. (2011). Research on cultural tourism in sacral and spiritual sites from the UNESCO world Heritage List. *International Journal of Heritage and Sustainable Development* 1(1): 49–59.

Harrison, Rodney. (2013). *Heritage: Critical Approaches*. London: Routledge.

ICOMOS. (2002). *International Cultural Tourism Charter: Principles and Guidelines for Managing Tourism At Places Of Cultural And Heritage Significance.* Paris: International Council on Monuments and Sites (ICOMOS).

Jadrnak, Jackie. (2015). Tribal Young People Stabilizing Bandelier Sites. *Albuquerque Journal*, 24 July 2015.

Keller, Robert H. and Michael F. Turek. (1999). *American Indians and National Parks*. Tucson: University of Arizona Press.

Kimball, Michael J. (2017). Heritage Place Building Theory, Heritage Impact Assessment and the Role of the Sacred Dimension. *Journal of Heritage Management* 2(1): 1–18.

King, Mary Ann. (2007). Co-Management or Contracting - Agreements between Native American Tribes and the U.S. National Park Service Pursuant to the 1994 Tribal Self-Governance Act. *Harvard Environmental Law Review* 31(2): 475–530.

King, Thomas F. (2013). *Cultural Resource Laws and Practice,* 4th edition. Walnut Creek, CA: AltaMira Press.

Kohler, Timothy A. (2004a). Introduction. In *Archaeology of Bandelier National Monument: Village Formation on the Pajarito Plateau, New Mexico,* edited by Timothy A. Kohler, pp. 1–17. Albuquerque: University of New Mexico Press.

Kohler, Timothy A., ed. (2004b). *Archaeology of Bandelier National Monument: Village Formation on the Pajarito Plateau, New Mexico*. Albuquerque: University of New Mexico Press.

Laxson, Joan D. (1991). How "We" see "Them:" Tourism and Native Americans. *Annals of Tourism Research* 18(3): 365–391.

Leshy, John. (2022a). America's Public Lands: A Sketch of Their Political History and Future Challenges. *Natural Resources Journal* 62: 341–360.

Leshy, John. (2022b). *Our Common Ground: A History of America's Public Lands*. New Haven: Yale University Press.

Liebmann, Matthew Joseph and Robert W. Preucel. (2007). The archaeology of the Pueblo Revolt and the formation of the modern Pueblo world. *Kiva* 73(2): 195–217.

Lott, Jason. (2020). *Superintendent's Compendium Bandelier National Monument*, National Monument, 2020, pp. 1–26. Washington D.C.: U.S. Department of the Interior, National Park Service.

Mamo, Duane, ed. (2021). *The Indigenous World 2021*. Copenhagen: International Work Group for Indigenous Affairs.

Mathien, Frances Joan. (2004). History of Archaeological Investigations on the Pajarito Plateau. In *Archaeology of Bandelier National Monument: Village Formation on the Pajarito Plateau, New Mexico,* edited by Timothy A. Kohler, pp. 69–116. Albuquerque: University of New Mexico Press.

Merlan, Thomas and Frances Levine. (2000). *Bandelier National Monument Study of Traditionally Associated Native American Communities: Ethnographic Overview and Assessment.* Washington, D.C.: U.S. Department of the Interior, National Park Service.

National Park Service. (NPS) (2015). *Bandelier National Monument*. Washington, D.C.: U.S. National Park Service, Department of the Interior.

National Park Service (NPS) (2016). *NPS Entering the 21st Century*. Washington, D.C.: U.S. National Park Service, U.S. Department of the Interior.

National Park Service (NPS) (2018). *Federal Historic Preservation Laws The Official Compilation of U.S. Cultural Heritage Statutes 5th edition.* State, Tribal, Local, Plans & Grants Division, Cultural Resources, Partnerships, and Science Directorate. Washington, D.C.: U.S. National Park Service, U.S. Department of the Interior.

National Park Service (NPS) (2020). *Cultural Demonstration Program*. Washington, D.C.: National Park Service, U.S. Department of the Interior.

National Park Service (NPS) (2021). *2021 National Park Visitor Spending Effects: Economic Contributions to Local Communities and the Nation.* . Washington, D.C.: U.S. National Park Service, U.S. Department of the Interior.

Neely, Sharlotte and Douglas W. Hume, eds. (2020). *Native Nations: The Survival of Indigenous Peoples,* 3rd edition. Vernon, BC, Canada: JCharlton Publishing Ltd.

New Mexico Tourism Department. (2018). *New Mexico Tourism Department 2018 Annual Report.* Santa Fe: State of New Mexico, 2018, pp. 1–27.

Ortiz, Alfonso. (2009). *The Tewa World: Space, Time, Being, and Becoming in a Pueblo Society*. Chicago: University of Chicago Press.

Ortman, Scott G. (2010). Evidence of a Mesa Verde Homeland for the Tewa Pueblos. In *Leaving Mesa Verde: Peril and Change in the Thirteenth-Century Southwest*, edited by Timothy A. Kohler, Mark D. Varien, Aaron M. Wright, pp. 222–261. Tucson: University of Arizona Press.

Ortman, Scott G. (2012). *Winds from the North: Tewa Origins and Historical Anthropology*. Salt Lake City: University of Utah Press.

Parsons, Elsie Clews, et al. (1996). *Pueblo Indian Religion*. Vol. 1. Lincoln: University of Nebraska Press.

Porter, Michael E., et al. (2002). *The Global Competitiveness Report 2001-2002: World Economic Forum, Geneva, Switzerland 2001*. New York: Oxford University Press.

Powers, Robert P. (2005). *The Peopling of Bandelier: New Insights from the Archaeology of the Pajarito Plateau*. Santa Fe: School for American Research Press.

Preucel, Robert, ed. (2002). *Archaeologies of the Pueblo Revolt: Identity, Meaning, and Renewal in the Pueblo World*. Albuquerque: University of New Mexico Press.

Richards, G. (2018). Cultural Tourism: A Review of Recent Research and Trends. *Journal of Hospitality and Tourism Management* 36: 12–21.

Richardson, Benjamin Barron. (2016). Interpreting Consultation: U.S. Fish & Wildlife Service and Native American Perceptions of Engagement Practices. M.Sc. Thesis, University of West Virginia, Morgantown, West Virginia.

Roberts, David. (2004). *The Pueblo Revolt: The Secret Rebellion that Drive the Spaniards out of the Southwest*. New York: Simon & Schuster Paperbacks.

Rodon, Thierry. (1998). Co-management and Self-determination in Nunavut. *Polar Geography* 22(2): 119–135.

Rothman, Hal. (1988). *Bandelier National Monument: An Administrative History*, Professional Papers No. 14, Southwest Cultural Resources Center. National Park Service, U.S. Department of the Interior.

Schillaci, Michael A. (2021). Another Look at Tewa Origins: The Biological Evidence. *Kiva* 87(2): 190–219.

Silverberg, Robert. (1970). *The Pueblo Revolt*. Lincoln: University of Nebraska Press.

Smith, Monica L. (2002). *The Historic Period at Bandelier National Monument*. Professional Paper No. 63, Intermountain Cultural Resources Management, 2002, pp. 1–97. Washington, D.C.: National Park Service, U.S. Department of the Interior.

Spence, Mark D. (1999). *Dispossessing the Wilderness: Indian Removal and the Making of the National Parks*. New York: Oxford University Press.

Stuart, David E. (1989). *The Magic of Bandelier*. Santa Fe: Ancient City Press.

Stuart, David E. (2009). *The Ancient Southwest: Chaco Canyon, Bandelier, and Mesa Verde*. Albuquerque: University of New Mexico Press.

Stuart, David E. (2010). *Pueblo Peoples on the Pajarito Plateau: Archaeology and Efficiency*. Albuquerque: University of New Mexico Press.

Sweet, Jill Drayson. (1981). *Tewa Ceremonial performances: the effects of tourism on an ancient pueblo Indian dance and music tradition*. PhD Dissertation, The University of New Mexico, Albuquerque, New Mexico.

Sweet, Jill D. (1991). "Let' em Loose": Pueblo Indian Management of Tourism. *American Indian Culture and Research Journal* 15(4): 59–74.

Sweet, Jill D. (2004). *Dances of the Tewa Pueblo Indians*. Second Edition. Albuquerque: University of New Mexico Press.

Timothy, Dallen J. and Stephen W. Boyd. (2006). Heritage Tourism in the 21st Century: Valued Traditions and New Perspectives. *Journal of Heritage Tourism* 1(1): 1–16.

Tinker, G. E. (1993). *Missionary Conquest: The Gospel and Native American Cultural Genocide.* Minneapolis: Fortress Press.

Toll, H. Wolcott (1995). *An Analysis of Variability and Condition of Cavate Structures in Bandelier National Monument*, Professional Paper No. 53, Intermountain Cultural Resources Center. Washington, D.C.: National Park Service, U.S. Department of the Interior.

Traylor, D., L. Hubell, N. Wood, and B. Fiedler (1990). *The 1977 La Mesa Fire study: an investigation of fire and fire suppression impact on cultural resources in Bandelier National Monument.* Southwest Cultural Resources Center Professional Paper 28. Santa Fe: National Park Service , Southwest Cultural Resources Center.

Two Bears, Davina (2019). Shimásání dóó shicheii bi'ólta' - My grandmother's and grandfather's school: the Old Leupp boarding school, a historic archaeological site on the Navajo reservation. PhD dissertation, Indiana University, Bloomington, Indiana.

UNESCO. (2003). *Convention for the Safeguarding of the Intangible Cultural Heritage.* Paris: United Nations Educational, Scientific, and Cultural Organization (UNESCO).

United Nations. (2007). *United Nations Declaration on the Rights of indigenous Peoples.* New York: United Nations.

United Nations Development Program. (2015). *United Nations Development Programme Social and Environmental Standards (SES) Guidance Note 4: Cultural Heritage.* New York: United Nations Development Program.

U.S. Department of Veterans Affairs. (2014). *Veterans' Preference and Special Hiring Authorities* M28R, Part VI, Section A, Chapter 8, July 2, 2014,.

Washburn, Kevin. (2021). Simple Tribal Co-Management: Using Existing Authority to Engage Tribal Nations in Co-Management of Federal Public Lands. *Iowa Legal Studies Research Paper 2021.* Iowa City: University of Iowa Law School.

Wennekers, Sander, et al. (2005). Nascent Entrepreneurship and the Level of Economic Development. *Small Business Economics* 24(3): 293–309.

Wilcox, Michael V. (2009). *The Pueblo Revolt and the Mythology of Conquest: An Indigenous Archaeology of Contact.* Berkeley and London: University of California Press.

Wilkinson, Charles. (2018). "At Bears Ears We Can Hear the Voices of Our Ancestors in Every Canyon and on Every Mesa Top": The Creation of the First Native National Monument. *Arizona State Law Journal* 50: 313–333.

World Bank and International Monetary Fund. (2011). *Global Monitoring Report 2011: Improving the Odds of Achieving the MDGs.* Washington D.C.: World Bank and International Monitoring Fund.

Yellow Horse, Aggie J., Nicholet A. Deschine Parkhurst, and Kimberly R. Huyser. (2020) COVID-19 in New Mexico Tribal Lands: Understanding the Role of Social Vulnerabilities and Historical Racisms. *Frontiers in Sociology* 5: 610355.

Zellmer, Sandra B. (1998). Indian Lands as Critical Habitat for Indian Nations and Endangered Species: Tribal Survival and Sovereignty Come First. *South Dakota Law Review* 43: 381–437.

Zeppel, H. (2006). *Indigenous E: Sustainable Development and Management.* Wallingford: CABI.

## *Chapter 9*

# COVID-19 and the Limits of Community-Based Ecotourism as a Sustainable Livelihood Diversification Strategy

## *The Case of the Indigenous Karo of Batu Katak, North Sumatra, Indonesia*

Roderick Taufan Jan Buiskool and Stasja Koot

The socioeconomic impact of the ongoing COVID-19 pandemic[1] is enormous; it affects social life and all economic sectors to different degrees. It often discloses existing inequalities, or increases them, with the brunt of the pandemic borne mostly by the poor and most vulnerable (Suryahadi, Al Izzati, and Suryadarma 2020). Arguably, the economic sector that has been hit hardest by the pandemic is the global tourism sector. Travel restrictions have had a dramatic impact on the industry, which was, according to the United Nations World Tourism Organization (UNWTO), the largest economic industry in the world. Tourism created one out of ten jobs (UNWTO, 2019a) prior to COVID-19. This situation has also affected Community-Based Ecotourism (CBE), which has been defined as travel to experience "natural" spaces and is intended to support both environmental protection and community development (Honey 2008). Local (often Indigenous) communities have a significant amount of control over the development and management of CBE, and a major percentage of the benefits is supposed to stay within the community (WWF 2001; TIES 2020).

Prior to the COVID-19 pandemic, CBET has often been presented as a sustainable livelihood diversification strategy and a way to increase the well-being of Indigenous local populations, often described as marginalized, living on the edge of or inside national parks (Ajake and Amalu 2012a; Amalu et al.

2017; Butler and Hinch 2007; Buiskool 2020; Carr, Ruhanen, and Whitford 2016; Grodach 2010; Leena et al. 2013; Wieckardt, Koot, and Karimasari 2022; Scoones 2015). Supposedly, CBET would also conserve nature and biodiversity, although this idea has also been contested, as has its potential to further develop the marginalized (Bratek, Devlin, and Simmons 2007; Hitchcock 2004; Koot 2016; Koot, van Beek, and Diemer 2016; Shultis and Heffner 2016; Wieckardt, Koot, and Karimasari 2022). Whereas some scholars refer to CBET as a shortcut to sustainable development (Loo & Davies 2006; Ma & Yeh 2010), others argue that CBET is overly reliant on market mechanisms and tends to have adverse social and environmental effects (Büscher et al. 2012; Büscher and Fletcher 2019; Castree 2003; Duffy 2008; Fletcher 2011; Higgins-Desbiolles 2009).

In this chapter, we investigate the effects of COVID-19 on CBET as a livelihood diversification strategy for Indigenous communities. We do this in Batu Katak, Sumatra, Indonesia. In particular, we cast light onto the pandemic's consequences for the Indigenous inhabitants of Batu Katak, who are mostly of the Karo Batak group. Batu Katak is situated adjacent to the Mount Leuser National Park (MLNP), one of the main biodiversity hotspots on Sumatra, consisting of a tropical forest and appealing to tourists for its natural scenery. Of Indonesia's fifty national parks, eleven are located on Sumatra (UNEP-WCMC, 2020). Despite their protected status, the country still has one of the highest rates of deforestation (McCarthy 2002), and Sumatra stands out due to intensive deforestation that has resulted in the conversion of 70 percent of the island's forested areas as of 2010 (Margono et al. 2012).

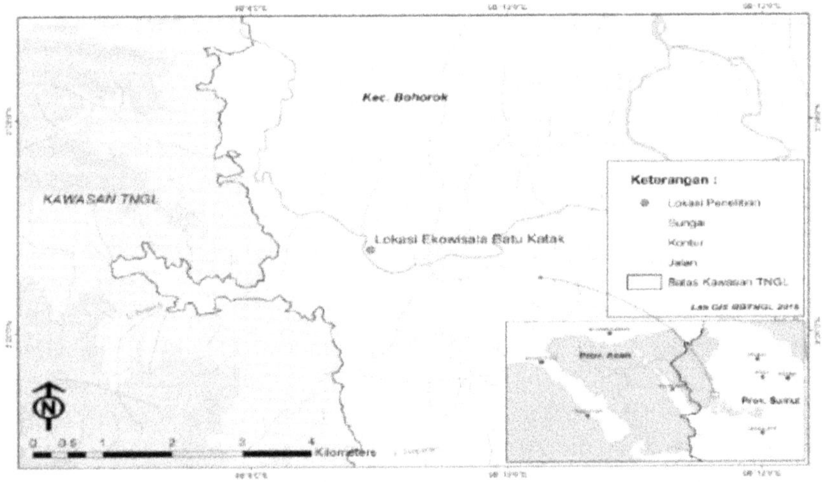


**Figure 9.1 Location of Batu Katak, Bordering the MLNP.** *Source*: Yuliana & Indarjo, 2018, 232.

As a crucial biodiversity area, the MLNP spans the North Sumatra and Aceh Provinces and is highly affected by deforestation, due to the increasing number of monoculture plantations, most importantly oil palm and rubber plantations. At the same time, MLNP is a high priority for biodiversity conservation (McCarthy 2002). The park covers 1,094,692 hectares with a population of approximately four million inhabitants within or adjacent to the national park (UNESCO 2015). Most people living in the vicinity of the MLNP can be classified as Forest Dependent Communities, as they are dependent on the natural capital from the forests for their sustenance and livelihood, making the conservation of the rainforest highly important for the local population (FAO 2017; UNESCO 2015). (See figure 9.1.)

As in so many other places, the COVID-19 pandemic started as a *shock* in Batu Katak but has developed into a typical *stress* (Chambers and Conway 1992). Shocks are sudden, unpredictable, and traumatic impacts, such as fires, floods, and epidemics, while stresses are continuous and cumulative pressures and therefore are to some extent predictable. Stresses include seasonal shortages, rising populations, or declining resources (Krantz 2001). For a livelihood to be sustainable, it is crucial that it is able to "cope with and recover from stresses and shocks" (Chambers and Conway 1992, 6). A livelihood's vulnerability results from a lack of resilience and an absence of the adaptive capacities required to respond to shocks and/or stresses. An important distinction here is between *environmental* sustainability, which refers to the external impact of a livelihood on local and global resources and other assets, and *social* sustainability, which concerns the internal capacity of a livelihood to withstand outside pressure in order to cope with shocks or stresses and to retain its ability to continue and improve over time, as explained above (Krantz 2001). In this chapter, the focus is on the social sustainability of the Karo in Batu Katak. The current situation that has arisen from the pandemic raises the important question of to what degree socially sustainable CBET can be regarded as a livelihood diversification strategy for Indigenous groups, who are often already in a vulnerable position relative to tourism (Carr, Ruhanen, and Whitford 2016; Koot 2016, 2023; Butler and Hinch 2007). This is also the case in Batu Katak (Wieckardt, Koot, and Karimasari 2022; Yuliana and Indarjo 2018).

Our chapter thus contributes to the understanding of ecotourism in anthropology by investigating CBET for an Indigenous group as a sustainable livelihood diversification strategy relative to the ongoing COVID-19 pandemic as a shock *and* as a stress. The chapter highlights the limits of CBET to preserve biodiversity and empower Indigenous people through tourism. The COVID-19 shock has had a significant impact on the livelihoods of the Karo in Batu Katak. Given that the international travel restrictions due to COVID-19 have not yet been lifted at the time of writing in November 2020 and have already lasted for around three-quarters of a year, the COVID-19 shock has effectively become a stress, affecting people's livelihoods in the long run. Many Karo employed in CBET have had to divert from CBET to other livelihood strategies, mainly back to agriculture in order to make ends meet. This has not only highlighted the vulnerability of CBET as a livelihood strategy for Indigenous people but also shows their dependence on CBE. While CBET is often considered a market-based mechanism that reduces marginalized groups' vulnerability, we argue that the ongoing COVID-19 pandemic has highlighted the vulnerability of CBET as a mechanism itself. This has important implications for the future of global efforts to preserve biodiversity and empower Indigenous people through tourism. Whereas ecotourism is often presented as an important livelihood diversification strategy to increase people's social sustainability, a shock/stress like the COVID-19 pandemic shows that it is important to keep approaching ecotourism as a *diversification* strategy, and not let it grow too dominant. However, due to various attractive benefits, ecotourism can tend to prevail as a livelihood strategy, increasing dependency on and vulnerability through CBET for Indigenous and other marginalized groups, as was disclosed in Batu Katak and its surroundings after the onset of COVID-19.

In the remainder of this chapter, we first present the case study of the Indigenous Karo and CBET in Batu Katak. Next, we present our results about the changes in the livelihoods of the Karo due to the COVID-19 pandemic, subsequently discussing the vulnerability of CBET as a livelihood diversification strategy and making comparisons with two other CBET hotspots nearby Batu Katak, namely Bukit Lawang and Tangkahan. Last, we conclude and reiterate our main argument.

## CASE STUDY: THE KARO IN BATU KATAK

### The Karo

The Karo are one of six Batak *suku* (which in Bahasa Indonesia translates to "tribe" or "ethnic group") with traditional homelands in North Sumatra, Indonesia (Kushnick 2006; 2010; Singarimbun 1975; Smith Kipp 1993). They are Indigenous to the peri-equatorial highlands of North Sumatra. Traditionally, the Karo are inhabitants of villages between 700 and 1,400 m (2,297 to 4,593 ft) above sea level. As subsistence and cash-cropping agriculturalists, the typical Karo household farms either wet (*sawah*) or dry (*ladang*) rice and other endemic and introduced vegetables, fruits, and spices, and engages in small-shop ownership, hunting, or other work (Kushnick 2006).

The area where the Karo live is referred to as *Taneh Karo*, which consists of the entire Karo Regency, the administrative unit just north of Lake Toba and west of the city of Medan (Kushnick 2010). A typical Karo village is arranged in a densely populated core, with fields and gardens on the periphery. The majority of Karo are Protestant or Catholic, though an increasing number are Muslim, and they speak an Austronesian language (*Bahasa Karo*) (Kushnick 2010; Smith Kipp 1993). Although today many have resettled elsewhere, the Karo villages in the *Taneh Karo* on or adjacent to the highland plateau tend to be largely ethnically homogeneous. The Karo who live in the surrounding areas of the Karo highlands are called *Karo Jahe*, derived from *enjahe*—which in the local Karonese language means "below"—referring to people living in the lowlands instead of the Karonese highlands.

Strictly speaking, it depends on one's definition of indigeneity whether the Karo can be regarded as Indigenous.[2] Although they are regularly described in such terms as Indigenous, native, or autochthonous (Slaats and Portier 1981; Smith Kipp 1993; Waddington 2002), including by the influential UNWTO (2019b), at other times there is no mention of such a status. Generally speaking, Indigenous people have often been described as people who are native to a region prior to the establishment of "modern" states (Carr, Ruhanen, and Whitford 2016; Hinch and Butler 1996; Sapignoli 2015). Such genealogical ancestry is a core characteristic of the global construct of indigeneity (as articulated by various institutions, including the UN), in addition to the characteristics of *non-dominance*, *cultural difference*, and *self-ascription* (Daes 1996; Koot 2023; Saugestad 2001; UNDRIP 2008).

To a degree, all four characteristics are applicable to the Karo, but some clearly more than others. In fact, under the influence of a long history of interactions with other groups and modernization, today the Karo have developed "plural identities." Still, however, their identities have a strong Indigenous component, which was aptly described by Smith Kipp (1993) as "indigenous disembledness" (10). She saw the plural identities of the Karo as "newly compartmentalized identities . . . attributing to modern societies (and especially to the exercise of modern forms of power)" (10). This form of self "has been there all along" and is "an indigenous variety that differs in kind from Western selves" (10). Indeed, indigeneity has always been a process of heterogeneous views, agendas, and identity politics, "and not a fixed state of being" (De la Cadena and Starn 2007, 11). It is important to realize that in Indonesia, despite being a signatory of the United Nations Declaration on the Rights of Indigenous Peoples (UNDRIP), government officials oppose the concept of "Indigenous peoples" and thus Indigenous activism that singles out particular communities deserving of special, "Indigenous" rights (Siringoringo and Mambor 2020; Tsing 2007). In fact, in most of Asia and Africa, Indigenous sovereignty—often at the command of Western imperialists—is seen as a threat to their precious and recently acquired national sovereignty (Tsing 2007). Today, Indonesians from *all* ethnic groups, and *sukus* whose ancestral roots lie mainly in the archipelago, are referred to as *Pribumi,* a word derived from the Sanskrit terms *pri* (before) and *bhumi* (earth), which translates to "first on the soil." The *Pribumi* are distinguished from Indonesians of known foreign descent, such as Chinese Indonesians, Arab Indonesians, and Indo-Europeans.

## Batu Katak

Batu Katak lies in the proximity of *Taneh Karo* and is a forest-edge village bordering MLNP, located in the Batu Jong-Jong village, Bahorok subdistrict, Langkat district, in the North Sumatra Province (see figure 9.1). The name *Batu Katak* is based on a local legend, in which a local woman found a toad-shaped rock while fishing in the Katak River. The toad was believed to hold magical powers. This power was believed to bring good fortune to the village, and thus its inhabitants named their village *Batu Katak*—meaning "Toad Rock"—in honor of the magical toad-shaped rock.

The population of Batu Katak totals 221 residents and consists of 60 households with 101 men and 120 women, made up primarily of Indigenous Karo (88 percent), Javanese migrants (10 percent), and a few Malay (2 percent) (Yuliana and Indarjo 2018). As Batu Katak is not located in the Karo highlands, the Karo who live in Batu Katak are called *Karo Jahe.*

As of 2020, there were fifty-two people aged less than sixteen years, and the working-age population between sixteen and sixty years consisted of 155 people (Yuliana and Indarjo 2018). During field research in April 2020, twenty-two people of working age—an increase from fourteen in 2018—were directly involved in CBET as official Batu Katak ecotourism guides from the *Lembaga Pariwisata Batu Katak* (LPBK; Batu Katak Tourism Organization) (Yuliana and Indarjo 2018). It is important to realize that, in Batu Katak, income from tourism is irregular, and that it is also sensitive because "working as a tour guide for one day can earn someone a wage that sometimes equals a week's work on a rubber plantation and is therefore a popular job" (Wieckardt, Koot, and Karimasari 2022, 2622). As such, ecotourism can also increase inequality within the village. Moreover, many residents are indirectly involved in CBE, as the shops, kiosks, restaurants, and hotels are frequented by tourists, providing an income to the residents who are not directly involved in CBE. Furthermore, as of 2020, the two Batu Katak residents who became civil servants worked as an elementary school teacher and a village midwife, and only two households were engaged in trade and owned food stalls.[3]

The LPBK was set up in 2010 and is fully responsible for ecotourism in Batu Katak. Since its inception, LPBK has been collaborating with the MLNP

Authorities (Yuliana and Indarjo 2018). As of 2020, the LPBK has fifty-seven members, with five divisions focused on specific areas of tourism in Batu Katak, namely "conservation," "guiding," "culture," "search and rescue," and "*pulungen tawar* (traditional Karo medicine)."

At the time of the study in April 2020, the average household income in Batu Jong-Jong village was around Indonesian Rupiah 2,150,000 per month (US$151.24). Agricultural production was still a mainstay in and around Batu Katak. As of 2020, fifty-six households worked in the agricultural sector, and some also worked as traders and guides as a side job. About half of the people from Batu Katak owned land around the village, most of which is located on the west side of the Berkail River, up to the border of the MLNP (Buiskool 2020; Wieckardt 2018). Those who did not own land worked as laborers on land owned by other people, either by tapping rubber or working in palm oil plantations.

Around Batu Katak, directly next to the village, is a karst forest, also called Gunung Kapur (Limestone Mountain). This forms a distinctive ecosystem in which the land is made up of limestone and creates a landscape that features vertical cliffs, sinkholes, underground streams, springs, and complex cave systems (National Geographic 2020). The Gunung Kapur, a community forest, plays an important role in the livelihoods of the residents; it covers around 210 hectares, and many residents of Batu Katak have cleared plots for small-scale plantations. Moreover, the community forest is also used for CBET purposes.

## Methodology

Batu Katak was chosen as the research site because it is known as a CBET initiative, where the local Indigenous community primarily manages the ecotourism. Ethnographic research was conducted in April 2020 for a period of three weeks, in the midst of the newly imposed travel restrictions. Ethnography, with its strong focus on "studying people as they live their day-to-day lives" (Emerson, Fretz, and Shaw 2011, 1), has proven particularly useful in studying *how* everyday lives, and thus the livelihoods of the Karo from Batu Katak, are influenced by CBE, as well as how the residents react to the possibilities and challenges it creates. The research was primarily conducted by the first author, who is a Dutch-Indonesian male, grew up in North Sumatra, and is a native Bahasa Indonesian speaker. Additionally, the second author conducted one exploratory interview with two tourism managers in Batu Katak in February 2017.

Among the local inhabitants of the Batu Katak village, eighteen CBET stakeholders were interviewed for the study. The participants included people directly involved in the CBET initiatives, such as tour guides, park rangers,

LPBK members, guest house owners, restaurant owners, kiosk owners, Civil Society Organization members, and members of the local government. Residents not directly involved in CBE—such as the local farmers in the village—were interviewed as well. This provided a balanced account of how ecotourism is perceived by people not directly involved in it. Participants were chosen based on the *snowballing* sampling strategy, where one respondent referred another respondent (Bryant 2014). In order to minimize the risk of having skewed data by only interviewing the more visible respondents who were pointed out by others, random sampling of the Batu Katak residents was also conducted throughout the study.

The methods of data collection were, first, in-depth semi-structured individual interviews in order to gain rich descriptions of people's experiences and thoughts (Roulston and Choi 2018). Semi-structured interviews include open-ended and more focused questions. They allowed the participants to freely converse about the research topic, thus providing research narratives and giving insights into the experiences of the participants (Roulston and Choi 2018, 233). Second, participant observation allowed us to gain insights into the activities of the CBET stakeholders and participants during activities such as jungle trekking (Watson and Till 2018). Throughout the research process, field notes were written in order to ensure a selective and systematic record and descriptions of activities in the field (Wolfinger 2002). The data collection was conducted in conjunction with a local NGO, the Sumatran Ranger Project. This allowed the first author to sample and contact the relevant CBET actors.

## COMMUNITY-BASED ECOTOURISM AND COVID-19 IN BATU KATAK

Tourism development in Batu Katak has been initiated by the local community, especially spearheaded by the LPBK members. According to the current head of the LPBK, CBET was seen as a means to increase the overall well-being of the local residents and secure resources for the future by reducing reliance on and problems associated with illegal logging. Additionally, two CBET villages in the area (namely Bukit Lawang and Tangkahan) further inspired the idea (see, e.g., Cochrane 1996; Janssens 2018; Ni'am, Koot, and Jongerden 2021; Rachmawati 2018). The initiators of CBET in Batu Katak, and most of the local population, were inspired by the positive influence CBET had on people's well-being in these other two destinations, and so they started CBET in Batu Katak in 2010. Prior to CBE, livelihood opportunities in Batu Katak were very limited, and many residents resorted to extractive livelihood strategies such as hunting and logging. Furthermore, some residents also relied solely on monoculture agriculture such as rubber and palm oil.

The combination of the Berkail River's panorama, the presence of endangered flora, and wildlife in the MLNP and in the Gunung Kapur makes Batu Katak an attractive place for visitors (Yuliana and Indarjo 2018), as well as an area with high conservation value. Most local guides reported that many CBET activities take place in the Gunung Kapur, and due to its rich biodiversity, there is a high chance of spotting wildlife. Specific attractions are some of the endemic flora, including the biggest flower in the world, namely the Titan Arum (*Amorphophallus titanium*) and Rafflesia (*Rafflesia lawangensis*) flowers. Moreover, endangered fauna such as the Sumatran orangutan (*Pongo abelii*), the Siamang or black-furred gibbon (*Hylobates syndactylus*), the Malayan sun bear (*Helarctos malayanus*), and the Sumatran tiger (*Panthera tigris sumatrensis*) can be spotted in or around MLNP. Other big charismatic animals, such as the Sumatran rhinoceros (*Dicerorhinus sumatrensis*) and the Sumatran elephant (*Elephas maximus sumatranus*) can still be found in and around the MLNP but not in the area of Batu Katak.

CBET activities in Batu Katak, MLNP, and the Gunung Kapur include jungle trekking, wildlife spotting, caving, bird watching, camping, and wild water rafting. The local guides can take guests for a hike for one or several days. Furthermore, tourists can enjoy Karo culture by viewing traditional Karo dance performances, joining cooking lessons including traditional Karo recipes, or visiting a *pulungen tawar* (traditional herbal medicine garden). Some of the main CBET supporting facilities in Batu Katak are coffee shops, food stalls, public toilets, a camping ground, and a public suspension bridge (Yuliana and Indarjo 2018). There are six kiosks and eight guesthouses in Batu Katak, all owned by local residents.

The total number of international tourists who visited Batu Katak in 2019 was 238, while the total number of domestic visitors from Indonesia, mostly from North Sumatra, was 20,356. This shows that besides international tourists, domestic tourism plays an important role in CBET operations in Batu Katak. Most of the activities such as trekking, rafting, or caving are usually done by international tourists, while domestic tourists tend to be more interested in day trips, picnics, and photo shoots at charming natural scenes (e.g., at a waterfall). This means that in terms of income, international tourists are economically much more interesting for the CBET stakeholders.

## COVID-19

At the time of the study in April 2020, Batu Katak was in complete lockdown due to national and international travel restrictions and stay-at-home orders. As a result, CBET operations had reduced significantly. As stated by a participant,

> We used to sell sweets, instant noodles, by the river bank in the centre of the village during the holiday's season [. . .] So right now we can see it during this COVID-19 period, no one can visit Batu Katak and it has reduced our income.

The initial shock of the pandemic immediately exposed the vulnerability of CBET as a sustainable livelihood strategy, in both the short and long term. According to an ecotourism guide, "Now with the pandemic young people see that there is no income from ecotourism, so maybe they are not so keen on working in the CBE sector." Thus far, in Batu Katak, the most important shock to CBET has been the COVID-19 pandemic. This has hampered the development of CBET and negatively impacted livelihoods, as all ecotourism operations were halted, causing a sharp decline in income. As reported by a resident, "Usually with tourists coming, I can work. Now there is no more work, and my husband can no longer work as a local guide either." Some participants also reported that they had lost their jobs. As one ecolodge staff member said, "I have been laid off for one month because there were no guests here because of COVID-19."

Given that the CBET initiative in Batu Katak is relatively young and in development, most people who work in tourism do so as a side job. The COVID-19 pandemic has reshaped the livelihood strategies of the residents of Batu Katak. As a local guide and guesthouse worker reported, "During the COVID-19 pandemic, the effect of no tourism on the livelihoods of Batu Katak [residents] is hugely impacting the resources for future well-being as well."

Many participants expressed that, due to the prolonged travel restrictions and lack of income from CBE, they have diverted their livelihood strategy (back) to agriculture. One participant said, "I can feel the influence of COVID-19 on tourism very much, and now I have started to mainly work on my small agriculture plot."

Additionally, many participants expressed that the (re-)diversification to agriculture feels like a "return" to how things were in Batu Katak prior to the start of CBE. A participant said that

> now because of COVID-19, there is really a big impact on the number of tourists coming [. . .] So it feels like our lives went back to how [they were] before CBE. Now we only work in our rubber plantations, and there is no extra income from tourism.

This emphasizes that income from CBET was used as a strategy for income diversification and is regarded as supplemental to most residents. However, another resident stated that "if this situation [lack of tourists] will persist, hunting and logging will definitely start again," since "the effect of reduced income from tourism on our livelihoods is huge, which will, in the long run, impact the resources for future well-being such as the conservation of forests and wildlife." This implies that, without the additional income from CBE, some residents might revert to practices such as logging and land clearing for agriculture and also to illegal hunting in order to make ends meet.

With these perspectives as a backdrop, the need for diversification of income has been highlighted. This was discussed by the chief of Batu Katak when he shared his own efforts to diversify: "I started with *trigona* [honey bee production], and it has been very helpful, as I can sell the honey. So far, the *trigona* [business] has been quite successful, and I can continue to get income even though there is COVID-19." Despite emphasizing the dramatic drop in income that has been an enormous challenge, many respondents also expressed hope that they will be able to earn income from CBET again in the long run. This was aptly stated by the head of Batu Katak, who said, "There's no way this COVID-19 pandemic will last forever, right? It will definitely end someday."

At the time of writing in November 2020, Batu Katak has been open for domestic tourists from Indonesia, meaning that there is currently income from CBE. Overall, however, the income from CBET has decreased seriously compared to the time prior to COVID-19 and the international travel restrictions. As mentioned above, the most profitable CBET activities are undertaken by international tourists.

## DISCUSSION

As explained by Chambers and Conway (1992, 6) regarding social sustainability, "a livelihood is sustainable when it can cope with and recover from stress and shocks [and] maintain or enhance its capabilities and assets." The national and international travel restrictions due to COVID-19 started out as a typical shock to CBET in Batu Katak, causing a sharp decline in the income of the residents. However, due to the ongoing international travel restrictions, the shock has now turned into a "stress," disclosing the limits of CBET as a sustainable livelihood diversification strategy for the residents of Batu Katak. This means CBET itself is rife with vulnerability while simultaneously often being presented as a market mechanism to improve resilience, vulnerability's counterpart. In the context of Batu Katak, this vulnerability shows because many residents have lost their livelihoods based on CBET due to COVID-19 and consequently have started to deploy various coping mechanisms, such as diverting their livelihood strategies from CBET to agriculture. This shows CBE's weak social sustainability to withstand external pressure and to cope with stress and shocks (Krantz 1999). Of course, CBET is not the only development strategy based on livelihood diversification for which this is applicable. Depending on the shock or stress at stake, all livelihood strategies have their limits. However, what is peculiar about ecotourism is that it is often presented as a silver bullet, but when one relies too much on only this strategy, vulnerability can increase, as a comparison with two other ecotourism destinations in the surroundings of Batu Katak shows.

## The Influence of COVID-19 on Surrounding Ecotourism Destinations

Batu Katak is situated around twenty-five kilometers from Bukit Lawang, which is known for the largest sanctuary of Sumatran orangutans and is one of the most popular tourism destinations in North Sumatra (Cochrane 1996). Furthermore, Batu Katak is not far (around fifty-six kilometers) from Tangkahan, another popular ecotourism destination in North Sumatra where tourists mostly go to engage with captive Sumatran elephants (Rachmawati 2018; Ni'am, Koot, and Jongerden 2021).

In Bukit Lawang, ecotourism has become the main source of income for most residents during the last few decades, while in Batu Katak it has thus far been only complementary to other livelihood strategies. In fact, pre-COVID-19 studies have demonstrated that, in Bukit Lawang, ecotourism has had a negative influence on agriculture, as land and food prices have increased while commodity prices for agricultural products have decreased (Janssen 2018).[4] Most of the people in Bukit Lawang are heavily dependent on ecotourism, especially international tourists, and are therefore hit particularly hard by the COVID-19 crisis (Serhadli 2020). While visiting Tangkahan in June 2020, several people reported that most residents previously working in ecotourism had then reverted to other livelihood strategies, mainly agriculture. This is relatively similar to what happened in Batu Katak. The comparison with Bukit Lawang and Tangkahan demonstrates the importance of developing truly *diversified* livelihoods, instead of relying on CBET as a dominant (or only) source of income, as is the case in Bukit Lawang. In the end, an important prerequisite of CBET is external and internal political, economic, and social stability. For instance, in 1998, Indonesian tourism experienced a significant shock during the Asian Financial crisis, and the ensuing public unrest impacted the tourism sector negatively (Henderson 1999). The same happened in the aftermath of the 2002 and 2005 Bali bombings in combination with SARS, the bird flu, and the Iraq war (Andriyanto 2020; Hitchcock and Darma Putra 2005). And again, closer to Batu Katak, shocks can be much more local: on November 2, 2003, flash floods occurred, which had a significant effect on the area and spotlighted the low resilience of CBET to a natural shock (BBC 2003). Ecotourism hotspot Bukit Lawang was hit hardest as the floods destroyed local tourist resorts and had a devastating impact on the local tourism industry in the area. Around 400 houses, 3 mosques, 8 bridges, 280 kiosks and food stalls, 35 hotels and guest houses, and the only irrigation system of the area were demolished by the flood (Janssen 2018). Furthermore, 239 people, including 5 tourists, were killed, and around 1,400 locals lost their homes (Lewis 2011; Janssen 2018). According to people in the area, including rangers and people from NGOs, the flash floods were due to illegal logging that occurred in the area (Aglionby 2003).

After the 2003 flood disaster, the Indonesian government closed Bukit Lawang for eight months in order to give the people time to recover. The (temporary) collapse of the tourism sector and the ensuing problems in the agricultural sector after the flash floods demonstrated that the area is very vulnerable to shocks related to the river (Janssen 2018). Bukit Lawang was not able to recover quickly, and some people left the area while others changed their livelihood strategy. Just like in Batu Katak currently, the local community in Bukit Lawang started to adapt their livelihood strategies to the new context in the area. However, given the magnitude of the floods in Bukit Lawang, many members of the community did not have housing or reliable access to food in the aftermath (Janssen 2018). Clearly, ecotourism in Bukit Lawang has been the victim of its own success, and today with COVID-19, the place is again hit much harder than Batu Katak, where thus far, CBET has not grown too dominant as a livelihood strategy.

## CONCLUSIONS

With COVID-19 hitting hard, 2020 has led to an enormous reduction of the number of tourists globally, including in Batu Katak. Through our examination of the pandemic's effects on CBE, and on the Indigenous Karo of Batu Katak specifically, we have investigated and shed light on the vulnerability of CBET itself for Indigenous and other marginalized groups as a sustainable livelihood diversification strategy. Whereas ecotourism is often presented as an important livelihood *diversification* strategy, it is important to regard CBET as a strategy to diversify income as opposed to becoming a *main* source of income. Due to various attractive benefits that come with CBE, including a higher income, ecotourism can also prevail as a livelihood strategy, increasing the dependency and vulnerability of Indigenous and other marginalized groups. This is evident in Bukit Lawang, where most people are highly invested in and dependent on income from ecotourism, and other options have consequently become limited. This shows the vulnerability of CBET as a livelihood strategy. The decline in income from ecotourism could pave the way for less sustainable practices such as illegal hunting, logging, clearing land adjacent to the protected areas, selling land, or cultivating monoculture plantations (such as oil palm and rubber). In the long run, these practices can degrade the natural environment and potentially worsen the situation of those already marginalized.

In Batu Katak, it is evident that the Indigenous Karo regard CBET as a complementary livelihood strategy, as it is still a developing initiative. However, under COVID-19, many Karo have had to divert to other livelihoods and mainly returned to working in the agricultural sector in order to make ends meet. This underlines the importance of having a diversified livelihood strategy and not being solely dependent on CBE, which is highly dependent on external political, economic, and social factors. We argue that the ongoing COVID-19 pandemic has highlighted the vulnerability of CBET as a mechanism itself. Whereas ecotourism is often presented as an important livelihood diversification strategy to increase people's social sustainability, a shock/stress like the COVID-19 pandemic shows that it is important to keep approaching CBET as a *diversification*, and not allow it to become the dominant, or even the only, strategy. However, due to various attractive benefits, ecotourism can have the tendency to prevail as a livelihood strategy, which tends to increase dependency on and vulnerability through CBET for Indigenous and other marginalized groups.

## NOTES

1. Coronavirus disease or COVID-19 is a contagious respiratory and vascular disease caused by severe acute respiratory syndrome coronavirus 2 (SARS-CoV-2), first identified in 2019 (Gössling, Scott, & Hall 2020).
2. Especially in anthropology, the term "Indigenous" has been subject to long debates, which go beyond the scope of this chapter, but see especially Kuper (2003) and special issues in, among other journals, *Anthropology Today* 2000: 16(4) and 2004: 20(2), *Social Anthropology* 2006: 14(1) and *Current Anthropology* 2004: 45(2) and 2006: 47(1). Barnard (2019) provides an overview of the debate.
3. See Yuliana & Indarjo, 2018, p. 234.
4. Wieckardt, Koot, and Karimasari (2022) also describe that some inhabitants of Batu Katak complain about increasing land prices, especially in relation to the Green Life nature conservation project, which also attracts many volunteer tourists.

## REFERENCES


Aglionby, John. "Flash Flooding Kills 72 in Sumatra Tourist Resort." *The Guardian*, November 4, 2003. https://www.theguardian.com/world/2003/nov/04/indonesia.johnaglionby.

Ajake, Anim O., and Titus E. Amalu "The Relevance of Tourism Development on the Economic Development of Cross River State, Nigeria." *Journal of Geography and Regional Planning* 5 (1) (2012): 14–20. doi:10.5897/JGRP11.122.

Amalu, T. E., E. I. Duluora, O. O. Otop, K. S. Emeana, and V. U. Omeje. "Assessment of Tourists' Patronage of Obudu Mountain Resort, Cross River State, Nigeria." *Journal of Hospitality and Management Tourism* 8 (4) (2017): 32–41.

Andriyanto, Heru "Pandemic Hits Bali's Tourism Harder Than the 2002 Attacks." October 12, 2020. Accessed November 25, 2020. https://jakartaglobe.id/lifestyle/pandemic-hits-balis-tourism-harder-than-the-2002-attacks.

Barnard, Alan. *Bushmen: Kalahari Hunter-Gatherers and Their Descendants*. Cambridge: Cambridge University Press, 2019.

Bratek, Oswald, Pat Devlin, and David Simmons.. "Conservation, Wildlife and Indigenous Tourism: Longhouse Communities in and Adjacent to Batang Ai National Park, Malaysia." In *Tourism and indigenous peoples: Issues and implications*, edited by Richard Butler and Tom Hinch, 141–57. Amsterdam: Butterworth-Heineman: 2007.

Bryant, Joanne. "Using Respondent-Driven Sampling With 'Hard to Reach' Marginalised Young People: Problems with Slow Recruitment and Small Network Size." *International Journal of Social Research Methodology* 17 (6) (2014): 599–611.

Buiskool, Roderick T. J. The Nexus Between Community-Based Ecotourism (CBE) and Well-being: A Case Study on Batu Katak Hamlet, Bordering the Mount Leuser National Park (MLNP). MA thesis, Maastricht University, 2020.

Büscher, Bram., & Fletcher, Robert. Towards Convivial Conservation. *Conservation and Society*, 17 (3 (2019)), 283. doi:10.4103/cs.cs_19_75

Büscher, Bram., Sullivan, S., Neves, K., Igoe, J., & Brockington, D. "Towards a Synthesized Critique of Neoliberal Biodiversity Conservation." *Capitalism Nature Socialism*, 23 (2) (2012), 4–30

Butler, Richard, and Tom Hinch, eds. *Tourism and Indigenous Peoples: Issues and Implications* (Amsterdam: Butterworth-Heineman, 2007).

Carr, Anna, Lisa Ruhanen, and Michelle Whitford. "Indigenous Peoples and Tourism: The Challenges and Opportunities for Sustainable Tourism." *Journal of Sustainable Tourism* 24 (8–9) (2016): 1067–79.

Castree, Noel. "Commodifying What Nature?" *Progress in Human Geography* 27 (2) (2003): 273–92. doi: 10.1191/0309132503ph428oa.

Chambers, Robert, and Gordon Conway. "Sustainable Rural Livelihoods: Practical Concepts for the 21st Century." *IDS Discussion Paper*. Brighton: Institute of Development Studies, 1992.

Cochrane, Janet. "The Sustainability of Ecotourism in Indonesia: Fact and Fiction." In *Environmental Change in South-East Asia: People, Politics and Sustainable Development*, edited by M. J. G. Parnwell and R. L. Bryant, 237–59, 1996.

Daes, Erica-Irene A. *Working Paper on the Concept of 'Indigenous People'*. Accessed December 18, 2020. https://digitallibrary.un.org/record/236429?ln=en.

De la Cadena, Marisol, and Orin Starn.. "Introduction." In *Indigenous Experience Today*, edited by Marisol de la Cadena and Orin Starn, 1–30. Oxford: Berg, 2007.

Duffy, Rosaleen. "Neoliberalising Nature: Global Networks and Ecotourism Development in Madagasgar." *Journal of Sustainable Tourism* 16 (3) (2008): 327–44.

Emerson, Robert. M., Fretz, Rachel. I., and Shaw, Linda. *Writing Ethnographic Fieldnotes*. Chicago: University of Chicago Press, 2011

Fletcher, Robert. "Sustaining Tourism, Sustaining Capitalism? The Tourism Industry's Role in Global Capitalist Expansion." *Tourism Geographies* 13(3) (2011): 443–61.

Food and Agricultural Organization (FAO). *Social Protection for Forest-Dependent Communities,* 1–15. Rome: FAO, 2017.

Gössling, Stefan, Daniel Scott, and C. Michael Hall. . "Pandemics, Tourism and Global Change: A Rapid Assessment of COVID-19." *Journal of Sustainable Tourism* 29 (1) (2020): 1–20.

Grodach, Carl. "Beyond Bilbao: Rethinking Flagship Cultural Development and Planning in Three California Cities." *Journal of Planning Education and Research* 29 (3) (2010): 353–66. doi: 10.1177/0739456X09354452.

Henderson, Joan C. "Asian Tourism and the Financial Indonesia and Thailand Compared." *Current Issues in Tourism* 2 (4) (1999): 294–303.

Higgins-Desbiolles, Freya. *Capitalist Globalisation, Corporatised Tourism and Their Alternatives*. New York: Nova Science, 2009.

Hinch, Tom, and Richard Butler. "Introduction: Revisiting Common Ground." In *Tourism and Indigenous Peoples: Issues and Implications*, edited by Richard Butler and Tom Hinch, 1–12. Amsterdam: Butterworth-Heineman, 2007.

Hitchcock, Michael, and Nyoman Darma Putra. "The Bali Bombings: Tourism Crisis Management and Conflict Avoidance." *Current Issues in Tourism* 8 (1) (2005): 62–76. doi:10.1080/13683500508668205.

Hitchcock, Robert K. "Natural Resource Management Among Kalahari San: Conflict and Cooperation." In *Indigenous Peoples' Rights in Southern Africa*, edited by Robert K. Hitchcock and Diana Vinding, 202–27. Copenhagen: International Work Group for Indigenous Affairs, 2004.

Honey, Martha. *Ecotourism and Sustainable Development: Who Owns Paradise?* 2nd ed. New York: Island Press, 2008.

Janssens, Elske. The Effects of Tourism on Agricultural Sustainable Livelihoods In Bukit Lawang and Timbang Jaya Indonesia. MS thesis, The University of Wageningen, 2018.

Koot, Stasja. "Contradictions of Capitalism in the Kalahari: Indigenous Bushmen, Their Brand and *Baasskap* in Tourism." *Journal of Sustainable Tourism,* 28 (8–9) (2016): 1211–26.

Koot, Stasja. "Articulations of Inferiority: From Pre-Colonial to Post-Colonial Paternalism in Tourism and Development Among the Indigenous Bushmen of Southern Africa." *History and Anthropology* 34 (2) (2023): 303–22.

Koot, Stasja, Walter van Beek, and Jeroen Diemer. "The Khwe of Namibia: Foragers between game, tourism and politics." *Anthropos* 111 (2) (2016): 497–511. doi:10.5771/0257-9774-2016-2-497.

Krantz, L. *The Sustainable Livelihood Approach to Poverty Reduction.* Stockholm, Sweden: Division of Policy and Socio Economic Analysis Swedish International Development Agency (IDA), 2001..

Kuper, Adam. "The Return of the Native." *Current Anthropology* 44 (3) (2003): 389–402..

Kushnick, Geoff. Parent-offspring conflict among the Karo of North Sumatra. PhD dissertation, Anthropology, University of Washington, Seattle, 2006.

Kushnick, Geoff. "Resource Competition and Reproduction in Karo Batak Villages." *Human Nature*, 21(1) (2010):62–81

Leena, K., Tuula, R., & Heikki, H. "Indicators of the Criteria for Good Participation in Ecotourism Planning at Local Level: A Nordic Case Study." *Journal of Sustainable Tourism*, 10(4) (2013):451–66.

Loo, T., & Davies, G. "Branding China: The Ultimate Challenge in Reputation Management?" *Corporate Reputation Review*, 9(3) (2006):198–210.

Lewis, Tim. "The story of Bukit Lawang." *Handcrafted Films.* March 04, 2011. Accessed November 17, 2022. http://handcraftedfilms.net/the-story-of-bukit-lawang/.

Ma, P. W., & Yeh, C. J. "Individual and Familial Factors Influencing the Educational and Career Plans of Chinese Immigrant Youths." *Career Development Quarterly*, 2(1) (2010), 32–43.

Margono, Belinda Arunarwati, Svetlana Turubanova, Ilona Zhuravleva, Peter Potapov, Alexandra Tyukavina, Alessandro Baccini, Scott Goetz, and Matthew C. Hansen. "Mapping and Monitoring Deforestation and Forest Degradation in Sumatra (Indonesia) Using Landsat Time Series Data Sets from 1990 to 2010." *Environmental Research Letters* 7 (3) (2012). doi:10.1088/1748-9326/7/3/034010.

McCarthy, J. "Power and Interest on Sumatra's Rainforest Frontier: Clientelist Coalitions, Illegal Logging and Conservation in the Alas Valley." *Journal of Southeast Asian Studies*, 33(1) (2002):77–106.

National Geographic. "Karst." In *National Geographic Resource Library, 2020.* Accessed May 31, 2020. https://www.nationalgeographic.org/encyclopedia/karst/

Ni'am, Lubabun, Stasja Koot and Joost Jongerden. "Selling Captive Nature: Lively Commodification, Elephant Encounters, and the Production of Value in Sumatran Ecotourism, Indonesia." *Geoforum*, 127 (018): 162–70. https://doi.org/10.1016/j.geoforum.2021.10.018

Rachmawati, Eva. "Tourism and Community Empowerment: Critical Insights from Indonesia." PhD thesis, Lincoln University, 2018.

Roulston, Kathryn, and Myungweon Choi. "Qualitative Interviews." In *The Sage Handbook of Qualitative Data Collection*, edited by Uwe Flick, 233–49. London: SAGE, 2018. doi:10.4135/9781526416070

Sapignoli, Maria. "Dispossession in the Age of Humanity: Human Rights, Citizenship, and Indigeneity in the Central Kalahari." *Anthropological Forum* 25(3) (2015): 285–305.

Saugestad, Sidsel. *The Inconvenient Indigenous: Remote Area Development in Botswana, Donor Assistance and the First People of the Kalahari*. Borås, Sweden: Nordic Africa Institute, 2001.

Scoones, Ian. *Sustainable Livelihoods and Rural Development*. Rugby, UK: Practical Action Publishing, 2015.

Serhadli, Sinan. "Market-Based Solutions Cannot Solely Fund Community-Level Conservation (Commentary)." *Mongabay,* May 12, 2020. Accessed May 31, 2020. https://news.mongabay.com/2020/05/market-based-solutions-cannot-solely-fund-community-level-conservation-commentary/. 2020

Shultis, John, and Susan Heffner. "Hegemonic and Emerging Concepts of Conservation: A Critical Examination of Barriers to Incorporating Indigenous Perspectives in Protected Area Conservation Policies and Practice." *Journal of Sustainable Tourism* 24(8–9) (2016):1227–42.

Singarimbun, Masri. *Kinship, Descent and Alliance among the Karo Batak*. Berkeley: University of California Press, 1975.

Siringoringo, Jakob, and Victor Mambor. "Indonesia." In *The Indigenous World 2020*, edited by Dwayne Mamo, 250–266. Copenhagen: International Work Group for Indigenous Affairs (IWGIA), 2020.

Slaats, Herman, and Maretta Portier. "Grondrechten en Zijn Verwerkelijking in de Karo Batakse Dorpssamenleving." University of Nijmegen, 1981.

Smith Kipp, Rita. *Dissociated Identities: Ethnicity, Religion, and Class in an Indonesian Society*. Ann Arbor: University of Michigan Press, 1993.

Suryahadi, Asep, Ridho Al Izzati, and Daniel Suryadarma. *The Impact of COVID-19 Outbreak on Poverty: An Estimation for Indonesia.* SMERU Working Paper, 2020 http://www.smeru.or.id/sites/default/files/publication/wp_covid19impact_draft.pdf.

TIES. "The International Ecotourism Society: What is Ecotourism?" *The International Ecotourism Society*, 2020. Accessed November 11, 2020. https://ecotourism.org/what-is-ecotourism/.

Waddington, Ray. "The Indigenous Karo Batak People." The Peoples of the World Foundation. 2002. Accessed November 11, 2020. http://peoplesoftheworld.org/text?people=Karo+Batak.

Tsing, Anna. "Indigenous Voice." In *Indigenous Experience Today*, edited by Marisol de la Cadena and Orin Starn, 33–67. Oxford: Berg, 2007.

UNDRIP. "United Nations Declaration on the Rights of Indigenous Peoples." New York: United Nations 2007https://www.un.org/development/desa/indigenouspeoples/wp-content/uploads/sites/19/2018/11/UNDRIP_E_web.pdf.

UNEP-WCMC. Protected Area Country Profile for Indonesia from the World Database of Protected Areas 2020. https://www.protectedplanet.net.

UNESCO. “Gunung Leuser.” *Ecological Sciences for Sustainable Development*. 2015. Accessed December 16 2019. http://www.unesco.org/new/en/natural-sciences/environment/ecological-sciences/biosphere-reserves/asia-and-the-pacific/indonesia/gunung-leuser.

UNWTO. “International Tourism Highlight: 2019 Edition.” Accessed 5 November. https://www.e-unwto.org/doi/pdf/10.18111/9789284421152.

UNWTO. Recommendations on Sustainable Development of Indigenous Tourism. Madrid: UNWTO, 2019.

Watson, Annette, and Karen E. Till. “Ethnography and Participant Observation.” *The SAGE Handbook of Qualitative Geography*, edited by Dydia DeLyser, Linda McDowell, Mike Crang, Steve Herbert, and Stuart Aitken, 121–37. London: SAGE, 2010.

Wieckardt, Chantal Elizabeth. From ‘Hidden Resistance’ to ‘Hidden Violence’: A Case Study into the Responses to the Green Life Project, North Sumatra, Indonesia, as a Form of Privatised Nature Conservation. MS thesis, The University of Wageningen, 2018.

Wieckardt, Chantal, Stasja Koot, and Nadya Karimasari. 2022. “Environmentality, Green Grabbing, and Neoliberal Conservation: The Ambiguous Role of Ecotourism in the Green Life Privatised Nature Reserve, Sumatra, Indonesia.” *Journal of Sustainable Tourism*, 30(11)(2022): 2614–30. doi: 10.1080/09669582.2020.1834564.

Wolfinger, N. (2002). “On Writing Fieldnotes: Collection Strategies and Background Expectancies.” *Qualitative Research*, 2(1) (2002):85–93.

World Wildlife Fund (WWF). “Community-Based Eco Tourism Community Interests Meet EnvironmentalResponsibility.”https://www.wwf.or.id/en/about_wwf/whatwedo/pds/social_development/communitybasedecotourism/ 2001.

Yuliana, Yayuk, and Slamet Indarjo. “Pengembangan Ekowisata Batu Katak Melalui Metode Swot Analisis.” *Talenta Conference Series: Agricultural and Natural Resources (ANR)* 1 (2) (2018): 231–37.

*Chapter 10*

# Change among the Wannilaeto of Sri Lanka

## *Land, Livelihoods, and Ecotourism Impacts*

Wiveca Stegeborn

The Wannilaeto (Veddahs) are the Indigenous people of the island of Sri Lanka, south of India.[1] Like hunters and gatherers in many parts of the world, global, national, and local processes affected their livelihoods and well-being significantly, particularly since Sri Lanka's independence in 1948. In the 1950s, settlers were brought to the Wannilaeto area from outside who engaged in commercial agriculture. Trees and non-timber forestry products were exploited intensively. This chapter addresses the challenges that have faced the Wannilaeto, including the implementation of the Accelerated Mahaweli Development Scheme, a large-scale irrigation agriculture project (1983–1992), which expanded competition for land; traditional forestland inhabited by Wannilaeto. A protected area, the Maduru Oya National Park, was established in 1983 on Wannilaeto land. Government troops moved in for military training during the latest Sri Lankan Civil War (1983–2009) and the Wannilaeto were displaced to government Rehabilitation Villages to become rice farmers. The coronavirus pandemic has affected ecotourism in Sri Lanka substantially, and the downturn in the economy and political corruption have impacted the Wannilaeto. The Wannilaeto today are coping with tourists, human trafficking, health challenges, and incorporation into the national economy as poorly paid workers. The forest dwellers, whose forest home was integral to their hunting-gathering economy, were considered intruders in their environment, and their lands are now largely in the hands of other people. This chapter examines the impacts of ecotourism on the Wannilaeto.

## DISPLACEMENT OF THE WANNILAETO

The Wannilaeto (Veddahs), the Indigenous people of Sri Lanka, were affected substantially by Sri Lankan government policy decisions, including the establishment of a large hydroelectric and irrigation project, the Accelerated Mahaweli Development Project (AMDP). This project came into existence because of an enduring trade imbalance between the much-needed rice imports from China and India and the decreased demand for raw materials such as rubber from Sri Lanka. The project's goal was to relieve the nation from external dependency on rice and to become self-reliant through agriculture. As the government wished to expand the energy-demanding industrial sector, they had to use water from the rivers to produce electric hydropower from their rivers instead of buying expensive oil (Dissanayake 1985). The Mahaweli Project hydro plans started in 1969 but did not affect the Wannilaeto until 1982, when their traditional forest had to be remodeled for intensive petrochemical agriculture. The Wannilaeto had to be extracted from their hunting grounds and displaced to Rehabilitation Villages inside the Mahaweli Development Project blueprint grid in Sri Lanka. "System A-H" covered the north and the eastern part of the island, reaching the southeast. Some Wannilaeto were placed within System C in the grid, not far from their traditional territory; others inside "System B," which was toward the northeast. Tunnels were blasted through mountains to lead water from rainforest rivers to artificial dams for hydroelectric power, and parts of the rainforest were strip-logged and bulldozed for rice-paddy cultivation. The rainforest was under great commotion. Elephants and other wildlife, humans, birds, iguanas, and insects lost their habitat.

To make the Wannilaeto move, government vehicles came to the forest settlements with coconuts, flour, kerosene oil, and hygienic articles such as scented talcum powder and Lux soap with a beautiful white beautiful woman on the wrapping paper. The talcum powder was to give a lighter facial complexion. The Wannilaeto were made to believe that it was going to be an everyday occurrence in their new villages.

The Wannilaeto were displaced into System B in the Polonnaruwa Province and into System C in the Uva Province. This displacement had negative impacts on their well-being since the new areas were not as rich ecologically as the previous ones.

Long curfews during the civil war between 1983 and 2009 among government soldiers and the Tamil Tiger guerrilla caused irreparable ruptures to the Wannilaeto social fabric. They could not visit or attend to their relatives' joys and sorrows or consult midwives, medicinal practitioners, and shamans. On the unlikely occasion, there was a Singhalese farmer around with a bull

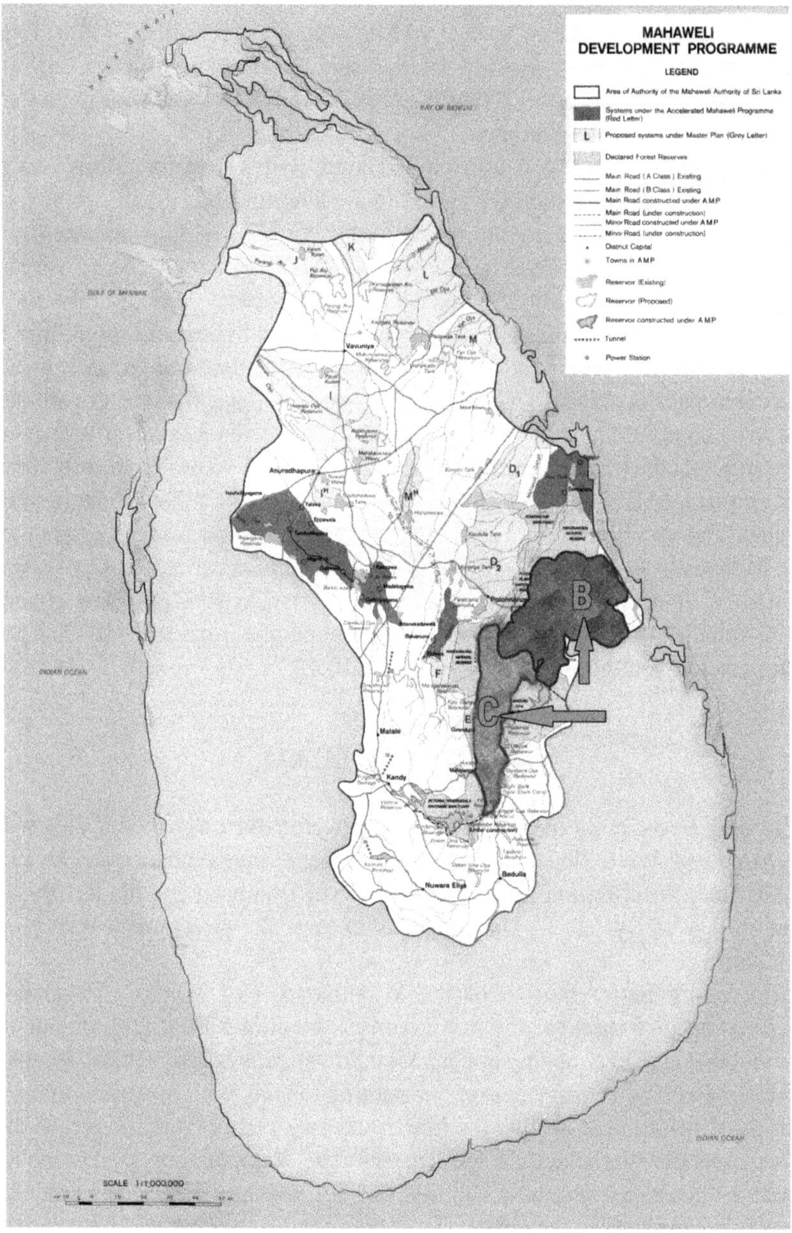


**Figure 10.1 Map of Sri Lanka with the Accelerated Mahaweli Development Project's Systems C & B.** *Source*: (Reworked from gov. Mahaweli Development Program) W. Stegeborn.

cart, a patient, for example, a woman in labor, could not be transported to the midwife from the other side of the national park barrier. The midwife, on the other hand, risked a fine of 10,000 Sri Lanka Rupees (about US$438 in 1983) and was detained if she crossed the park border to attend to a delivery at a patient's house (Daily-News-Reporter 1998). They all risked heavy fines and persecution by the government.

The saying: "The Road to Hell is paved with Good Intentions," is no exception here. The topic of a "good life" came up occasionally in government discussions regarding the future of the Wannilaeto. One argument was that the Wannilaeto should be allowed to make their own choice whether they wanted to adapt to a Singhalese/Tamil sedentary agricultural way of life or not. The government argued that hunters and gatherers should not be prevented from following a Western lifestyle with advanced technology and a secure material standard. Despite the Wannilaeto appeal to the government not to be displaced, and the response letter from President Jayawardena assuring them that they were allowed to remain where they were (Weerakon 1982), government AMDP officials continued to campaign for a displacement to the Rehabilitation Villages. Once the people moved to their respective systems, the Mahaweli Authorities said, the foragers would receive free plant seeds, fertilizers, flour, lentils, soap, and talcum powder. In 1982, the first government vehicles transported one family compound after another with their pots and pans to Systems C and B. (See figure 10.1.)

## THE GRAND SHAM

Returning to the point of free choice, the government was asked if they were prepared to listen to the Wannilaeto. There were arrangements made in 1983 to provide transportation for the Wannilaeto of Dambana and the Rehabilitation Village in System C, Henannegala, to meet the government authorities in Colombo.

The ailing representative of the Wannilaeto, Uru Warige, Tissahamy, received an invitation to stay in a luxurious hotel in Mount Lavinia outside Colombo. There, Tissahamy and his kindred would meet with Prime Minister Premadhasa. The primary reason for wanting to meet with the prime minister was the displacement despite the March 23, 1982, letter from the president's office to let the Wannilaeto be left in peace. The Wannilaeto wished to inquire about the government letter's validity. Besides, the ones displaced by the systems were guaranteed free plant seeds, flour, and so on, but had not received more than a few parts of the guaranteed commodities after they had moved. Those, and other issues were on Tissahamy's mind as the patriarch gathered a group of people and got ready to go to the meeting.

At the date of the appointment, the members of the Wannilaeto delegation were picked up by a vehicle to make the ten-hour journey across the country to the capital. In the hotel, the five to six-person delegation was crammed into one small room and they were not allowed to walk outside. Hungry and thirsty, they asked for the prime minister, only to hear that they would meet him later. Suddenly, they were taken out of the room and guided to a stage with bright, shining spotlights. Blinded at first, they could not see, but below, in the dark, were hundreds of people brimming with expectancy to see them. The Wannilaeto were asked to dance, sing, and perform spirit possession ceremonies in front of the high-paying public. The foragers did what was expected as they were told they would meet the prime minister afterward. Having met the demand, they were returned to the room where they were fed and brought to the vehicle to be driven back to the village. The whole thing was a sham. They never met the prime minister.

## DROUGHTS AND DISEASE

With the lack of rainforest, the weather pattern changed, and the seasonal rains became irregular and scanty. There were droughts. The water canals dried out. The harvests were poor, despite government-sponsored petrochemical fertilizers. The soil became sterile, and the groundwater was contaminated with chemicals, arsenic, and cadmium among other artificial agents (Jayasumana et al. 2014). People fell sick from chronic kidney disease (CKD), a kidney disease of unknown etiology according to government research (Rajapakse et al. 2016; Wanigasuriya et al. 2011). In 2023, half of the households in the rehabilitation village, Henenegala, in System C, had lost at least one family member to the disease.

There were causes related to agricultural practices, geographical distribution based on the prevalence and incidence of CKD, and contaminants identified in drinking water. The link between agrochemicals or heavy metals and CKD remains to be established, said faculty members of health institutions in Sri Lanka (Rajapakse et al. 2016). This issue still needs to be investigated by a non-governmental independent party.

## THE FOURTH PRESIDENT'S PROMISE

With leftover islets of trees parceled up between roads trafficked by heavy machinery, flat worksites, and noisy constructions to build a new town, Girandurukotte, the animals fled despairingly. Nature conservation agencies and international environmental organizations advised the government to leave the remaining plots of nature alone. The government could obtain revenues from national parks with tourist recreation instead. This would give plenty of job opportunities for jeep safaris, lodging, and guided tours in different languages.

The Maduru Oya National Park was designated on November 9, 1983, to "protect and preserve indigenous natural resources and ecological processes" and to "provide for the enjoyment of park resources through opportunities for visitor use" (Mahweli-Environmental-Project 1987:7). The national park is 58,850 hectares (227.2 sq. miles) in size. The Wannilaeto forest habitat became a recreational park for paying tourists. The money was collected by the government and not shared with local communities. There were information and interpretative services, protection, comfortable accommodation, and access to recreational activities for the visitors. The Environmental Program provided transportation to facilitate circulation within the park. Information was provided partly by trained trackers and guides, none of whom were Wannilaeto, who accompanied the tourists on their tours. The Master Plan for the Maduru Oya National Park suggested that an outdoor exhibit would describe the forest people that once lived there. What was not exhibited was why they were not still there, that foragers and forest fauna were relocated elsewhere. No board explained that they had to give way to both local and international visitors, that the Wannilaeto and their forest had become a recreational park where the trees served as an ecological lung for government dam projects and that the Wannilaeto left in the buffer zone were objects of interest to the visitors.

## HUNTERS AND GATHERERS, NO MORE

Signs were planned to depict flora, fauna, and cultural objects at different points throughout the park. The story of the Wannilaeto would be shown in displays exhibiting their implements, weapons, and ways of life (Mahweli-Environmental-Project 1987:44). The blanketing of the Wannilaeto was total. The outdoor museum displays gave the impression that they no longer existed, that the Wannilaeto were long gone, to the point of no return. By putting them in a distant past, it made the visitors believe that the forest dwellers had disappeared into the haze of long-gone history. In the Rehabilitation Villages, they were given new names, in Singhalese; they had to follow a new religion—Buddhism, and they were taught the history of their conquerors. Their past was eradicated. This way, the Wannilaeto would not get rebellious ideas of disobedience. (See figure 10.2.)

**Figure 10.2 Uru Warige, Sudda Preparing a Chew of Betel.** *Source*: W. Stegeborn.

## MASS TOURISM

The tourist industry developed from a "researchers only" policy to mass tourism in air-conditioned buses from the tourist resorts near the international airport. A World Tourism Organization study of mass tourism, which includes sex tourism and ecotourism in "developing" countries, reveals that "when a country considers tourism as a pillar of its economic development, it does so at the expense of human dignity" (Singh et al. 1989:34). Frank Go, Head of Tourism Marketing in the Department of Marketing Management at Rotterdam Erasmus University (RSM), continues, "it ultimately leads to the violation of human rights" (Ibid). Sri Lanka Tourism Board is firmly committed to transforming Sri Lanka into Asia's foremost tourism destination (Government-of-Sri-Lanka 2011). The main purpose for their visit was "pleasure," and the main consumers were men (Government-of-Sri-Lanka 2008).

## ECO-ETHNO-TOURISM

With time, global awareness of climate change and the importance of biodiversity, ecotourism, and ethno-tourism evolved. Anthropologists West et al. state,

> If there is a protected area, some form of ecotourism likely uses it, and if ecotourism enterprises are present, some protected areas likely exist in the vicinity. Because of this connection, people living in and around protected areas interact with ecotourism as a revenue source, as a set of social relationships that bring nature and culture to areas where they did not exist before, and as a conduit for visitors from other places. It brings new ways of seeing and using people's surroundings to already existing socioecological landscapes and creates new boundaries. (West et al. 2006:262)

West et al. explain that ecotourism causes conflict between people who have money because of tourism and those who do not;

> it created images of people living in the park as indigenous and those moved outside of it as villagers; and it created a dependency on conservation for jobs and income. (refs in text, Rugendyke & Son 2005) (West et al. 2006:262)

In the buffer zone village of Dambana, this is a reality. Many of the Wannilaeto outside the Maduru Oya National Park are economically dependent on tourism. The Wannilaeto, national and international entrepreneurs lobby hotels and tourist agencies to bring visitors to their private compounds. This nature and human

**Figure 10.3 U. W. Sita Dammayanti Cuts Straw to Thatch Roofs.** *Source*: W. Stegeborn.

consumption is tempting revenue, not just for the government and the Sri Lanka Tourism Development Authority but also for the Wannilaeto. (See figure 10.3.)

After the ban on Wannilaeto means of subsistence, on November 9, 1983, the Wannilaeto were branded "criminals" according to the *Sri Lanka Fauna and Flora Protection Ordinance of 1937*. It forbade them to enter the protected area, now called national park, for any purpose other than "observing the fauna and flora" (Government 1937:XVII/297, art. 3 b). No animal should be trapped or hunted (Government 1937:XVII/297, art. 3 c). No reptile or bird eggs should be collected, and no fish taken (Government 1937:XVII/298, art. 3 d). No one was allowed to carry a gun or other explosives, such as a muzzle-loader (this was also part of the Firearms Ordinance) nor carry a kindle or fire, clear land, or erect a building (Government 1937: XVII/299, art. 4 f).

Except for researchers and people with special permission, only the director for Wildlife Conservation and park staff were allowed inside. The former hunters and gatherers could not go to their forest settlements to collect their kitchen garden chilies, manioc, beans, or potatoes. The law had to be indiscriminately followed by all people, international and Sri Lankans alike; Tamils, Sinhalese, Moors, rich and poor, and the Wannilaeto. Park guards were placed in the forest to see that the law was followed.

## SEEKING JUSTICE AT THE UNITED NATIONS

Unconvinced by presidential promises, the Wannilaeto once more sought help from international human rights organizations outside their country. They formed a delegation to make a second attempt to attend the UN Working Group for Indigenous Populations in Geneva, Switzerland, in 1996 (Stegeborn, et al. 1996). The first attempt was eleven years earlier in 1985 when a cabinet decision (IWGIA 1985) prevented them from leaving the country. This time, however, in 1996, they managed to get out of Sri Lanka, and they presented their case to the UN. As a result of the international community's reaction, putting pressure on the Sri Lankan government, president number four, counting from the time the Wannilaeto were displaced in 1982, assured the following statement in 1998,

> The declaration by me, Her Excellency Chandrika Bandaranayake Kumaratunga—The President of the Democratic Socialist Republic of Sri Lanka. I hereby declare for the information of the people of Sri Lanka that I will offer opportunity to the Veddha people or Wanniyale Eththo to continue their traditional way of life (if they like to do so) and to make progress of their civilization: . . . (Bandaranayake Kumaratunga 1998:38)

**Figure 10.4 1996 UN, Geneva. U. W. Wanniya, U. P. W. Sudubandia and U. W. Punchi Banda.** *Source*: Stegeborn.

Consequently, the Environment and Natural Resources Ministry was instructed to issue special Maduru Oya National Park identity cards (MAOY-NP ID) for the Wannilaeto to avoid the quandary of mistaken identities. The card proved the authenticity of the Wannilaeto in case they met a warden inside the park. Twenty Maduru Oya National Park (MAOY-NP) cards were issued, permitting the Wannilaeto to carry out their traditional subsistence in the forest park. More would come in August, the government promised (Wijayasiri 1998a, b).

The President's leniency toward the Indigenous people did not agree with the majority population. People were openly hostile and expressed their anger in person and in newspaper articles. The old Wannilaeto spokesman, Uru Warige, Tissahamy, was especially targeted (Niyangoda 1998a, b; Rafiq 1999; Wijesinhe 1999). (See figure 10.4.)

## PUNCHI BANDA'S STORY

Maybe it was a blessing, at that time, for the leader Uru Warige, Tissahamy, to bid farewell to his wife, sons, daughters, and grandchildren to join his

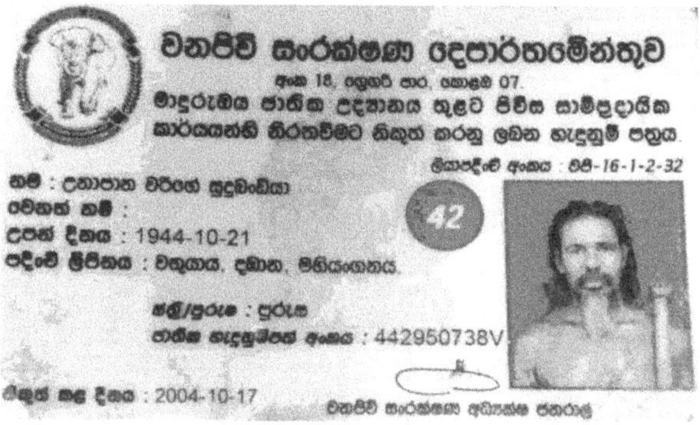


**Figure 10.5 U. P. W. Sudubandia's Maduru Oya National Park ID card.** *Source*: W. Stegeborn.

ancestors at the holy mountain Unukirigala on May 29, 1998. A few months after Tissahamy's passing, Uru Warige Punchi Banda, thirty-six years old and a father of five, went with his two relatives to hunt deer inside the Maduru Oya National Park. Suddenly, Punchi Banda was shot in the back by a national park guard (Lankapuvath-Correspondence 1999). Afraid for their lives, the other men escaped, but Punchi Banda fell and was caught and brought to the closest hospital in the area. The wardens took the dead game. (See figure 10.5.)

Although Punchi Banda did not yet have a printout of the national park card, he was hunting with the mistaken belief that he had a traditional right to do so as assured by the president. The bullet penetrated his spine, and he became paralyzed from the waist to his foot on one side (Rafiq 1999). Not knowing where Punchi Banda was taken, his wife Karunavatti and children began searching from town to town for their missing husband and father in hospitals, prisons, and morgues. They inquired of the local Director of Wildlife about the whereabouts of their family member, and they were told that he was at a nearby hospital in Girandurukotte. When the family arrived at the hospital, Punchi Banda was gone. The personnel understood the gravity of Punchi Banda's medical condition and had sent him to the regional hospital in the provincial capital, Badulla. With efforts from relatives and friends to finance the search trip, Punchi Banda's mother, his wife, and children traveled to Badulla to see their father/son/husband, but he had already been transferred to the General Hospital ward thirty-three in the nation's capital, Colombo.

Paralyzed, drooling, and incapacitated to speak because of facial palsy, the family found him foot-cuffed to his bed, guarded by two jailors in case he ran away. With tears in his eyes, he finally saw his family. His old mother prayed beside his bed. At that stage, he had not yet had a catheter inserted, nor did he have a colostomy bag on his stomach. (See figure 10.6.)

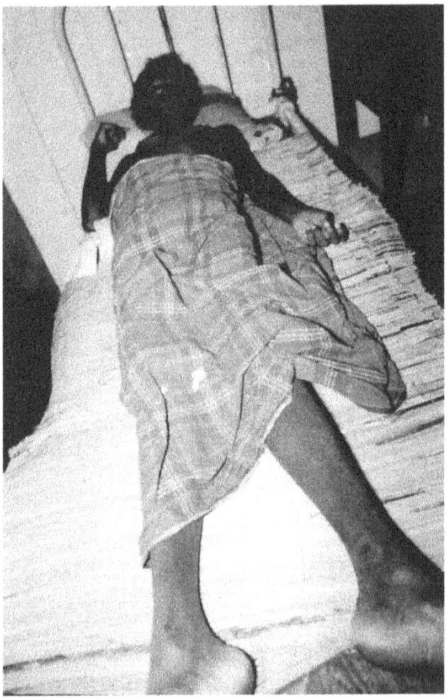

**Figure 10.6 Uru Warige, Punchi Banda Paralyzed in Hospital Bed.** *Source*: W. Stegeborn.

He was subsequently moved to a new ward reserved for spine surgery. There was a chance that Punchi Banda could get 50 percent better if he went through six months to a year of physiotherapy at a Rehabilitation Clinic outside Colombo.

## CAVE-MAN THINKING

Most Sri Lankans are unaware of how the Wannilaeto people live in modern times. Many believe they dwell in caves, hunt with bows and arrows, and apart from the loincloth, are naked. These images come from India's epics in the Ramayana, with its deeds and adventures among Gods and deities; the Mahavamsa with the Singhalese glorious conquest of Sri Lanka's demons and evil spirits; and Robert Knox's descriptions and drawings from novels and documentarists and folktales (Knox [1681] 1911; Madahapola-Ratemahatmaya 1973; Punchihewa 1989; Spittel 1957; Spittel 1950; Wijesekera 1964).

Based on these views, the park guard who shot Punchi Banda made a crafty twist. He said he targeted Punchi Banda because he "mistook the Vedda for a terrorist as he was wearing blue jeans" (Rafiq 1999). Consequently, he, the warden, charged Punchi Banda with attempted murder on December 23, 1998 (Arun Tampoe 1999).

## WHEELS OF JUSTICE

On February 26 (1999) Punchi Banda was asked to present himself at the Dehiattakandiya Magistrate's Court in System C. There were two charges against him:

1. Wildlife case no. (AR 2256/1998) under the Flora and Fauna Act for trespassing the park border (Arun Tampoe 1999).
2. Police case no. (1453 B report under section 300 Penal Code) for attempted murder (Arun Tampoe 1999).

Punchi Banda could not attend because of his paralysis. A certified letter of his medical condition was delivered to the judge, followed by an explanation of why the patient could not present himself (Arun Tampoe 1999). Displeased voices in the courtroom mumbled that there should be additional charges for not appearing. In Colombo, Punchi Banda's assigned Singhalese defense lawyer had not cared to contact Punchi Banda. He was unaware of his client's health condition since he had not gone to see him. The day before the court, the lawyer phoned and said that he was unable to appear at court for Punchi Banda due to "family matters" adding conspiratorially that there were "several cases pending in which Veddahs are involved in harassing the Wildlife Department. officers" (Arun Tampoe 1999). One may ask whether Punchi Banda's defense lawyer carried out a foregone conclusion of his case.

With no defense and an absent client, the prosecution could not proceed. Members of the court and interested parties, however, did address the local wildlife director regarding the incident. They questioned in which way Punchi Banda was a menace to the park guard, as he ran in front of him carrying his shotgun with the barrel forward. If Punchi Banda threatened the guard and the guard shot Punchi Banda in self-defense, the Wannilaeto man ought to have been shot in the stomach, not in the back. With an air of arrogance, the local wildlife director refused to reply. Pressured by common sense and convincing arguments, the local wildlife director realized that his story would not hold, and Punchi Banda was released on 1,000 Rs. bail even though bail for trespassing the park border usually cost more (Arun Tampoe 1999). The attempted murder allegation remained.

## WELIKADA HIGH-SECURITY PRISON

Punchi Banda's paralysis aggravated after the spine surgery in Colombo. He could end up in a critical condition if more nerves inside the vertebral column were disconnected. Half of his face became immobile, and his bladder ceased to function. He had to pass urine into a bag through a catheter and was shortly thereafter opened again to have a colostomy bag attached to his colon. Both legs became paralyzed, and his general prognosis was dim.

Based on the park warden's allegations that Punchi Banda had attacked him to kill, Punchi Banda was rearrested and brought to a high-security prison, Welikada, outside Colombo. Welikada is as infamous as San Quentin or Folsom State Prison in the United States. It is the most notorious prison in Sri Lanka, known to hold hardened criminals, serial murderers, and rapists, and it is ill-famed for its prison massacre in 1983.

Punchi Banda was dying, deteriorating by the day. At this point, he lay on a dirty straw mattress drenched with urine on the cold prison floor. He endured one month. On Friday, April 30, 1999, the Forest Spirits released him from the pain in his dysfunctional body, and he entered a merciful final slumber.

## MINISTER OF LEGAL AND PRISON REFORMS VISITS "VEDDAH" SHOW

In a previous segment, it was depicted that the Wannilaeto were compelled to enact dances on stage, feigning possession by ancestral spirits, under the assurance of a subsequent meeting with the Prime Minister. That was not

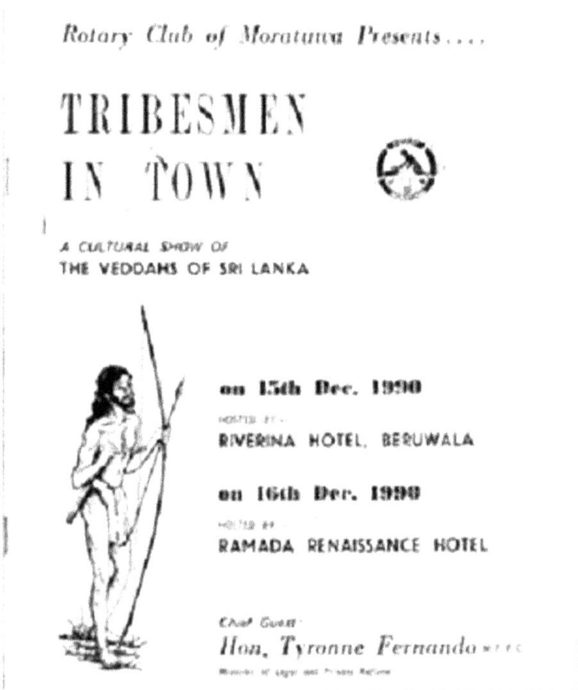


Bottom text reads:
Minister of Legal and Prison Reforms

**Figure 10.7 Tribesmen in Town Arranged by the Rotary Club of Moratuwa (the Town).**
*Source*: W. Stegeborn.

the only occasion. The gentlemen's club Rotary of Moratuwa arranged one exhibit which was extraordinary since the chief guest was the minister of Legal and Prison Reforms, the man responsible for Welikada Prison where Punchi Banda died. This was a masterpiece of awkwardness.

The Rotary Club and the executives set up a show in the most luxurious hotel in the heart of Colombo, the Ramada Renaissance Hotel. The guests were thrilled to see the Wannilaeto. The Rotary Club president promoted the sale of illegal forest products to tourists made by the forest people (Rotary-Club-Moratuwa 1990). The vending of national park resources is against the law but is viewed leniently when it involves the government. Contrary to the expected, the minister of Legal and Prison Reforms did not appear troubled by the presence of forest products or by the persecution of the Wannilaeto. Instead, he wished Rotary Moratuwa "all success" (Rotary-Club-Moratuwa 1990:2, message from chief guest). In the Welcome brochure *Tribesmen in Town*, there is a description of the Indigenous forest dwellers of Sri Lanka: (See figure 10.8.)

**Figure 10.8 Rotary Promotes Ivory and Forest Products from the National Park. The Illustration Comes from a Rice Paper Sheet the Rotary Club at Moratuwa Printed for the Exhibit.** *Source*: W. Stegeborn.

> Their way of life is different to that of ours. To be precise, their's [*sic*] is *not* a *civilized* way of life. In the days gone by and even now, they are traditionally hunters and *eat the bark* of the riti *tree*. The Veddahs' are *dark* in *complexion* [*sic*] with *flat noses*. They speak a different dialect. They are scantily dressed. They inhabit the jungles of wanni (Rotary-Club-Moratuwa 1990:5) (See figure 10.9.)

The legal incongruity was perplexing. Had the Welikada Prison minister seen the Wannilaeto the day before he allowed the dying hunter into his high-security Welikada Prison, he may have refrained from receiving illicit park products from Punchi Banda's relatives by encouraging the participants to buy bottled bee honey, medicinal plants, and animal teeth and claws during the exhibition. In some countries, this is a felony called A*ccessory of a Crime*. At the show, however, the Wannilaeto were as much under impunity as the minister himself. In the brochure "Tribesmen in Town," there was a description of the "Veddahs for the tourists." The purpose was to enlighten the spectators about the nation's Indigenous people.

**Figure 10.9 Passengers in Tourist Buses and Three-wheeler Taxis Come to Look at the Wannilaeto People in 2023.** *Source*: W. Stegeborn.

## THE SHOW MUST GO ON

Unaware of the pamphlet maker's made-up *platyrrhine* features of the Wannilaeto and their peculiar food habits, the rainforest people initiated the show. They presented, good-naturedly, a bow and arrow as gifts to the honored guest, the minister of Legal and Prison Reforms, as the custom bids in Sri Lanka. It was a rather sinister situation where everyone pretended except the duped spectators. It should be added that some of the international tourists left the room in the middle of the show, discomforted by what they saw.

Sometimes the Wannilaeto had to act as if they were possessed twice a day. It took much of their strength and concentration. They were exploited in several ways, all of them based on the presumed customs of their culture: balanced reciprocity, animistic beliefs, traditions of dance, myth, and material life. Reduced to fairytale malignant figures, they acted and sold themselves as those characters to the spectators. The visitors were about to witness a unique experience of "primitive people" possessed by evil spirits. Everyone but the tourists knew that the performance was faked, but the show gave validity to the visitors through deception. As time went by, the morale of the Wannilaeto men deteriorated and the show became a travesty of themselves. To perform night and day, some of the younger generation started to drink before the shows. I witnessed occasions where my dear ones were completely drunk, jumping like frogs all around the hotel, shouting and screaming and laughing as if possessed by spirits to fool the onlookers. The Wannilaeto could act in any way they wanted since tourists had no idea about spirit possession. It was not a nice sight. It became a mockery of themselves and the audience for believing this was for real. Instead of enhancing their mindful and beautiful culture, they dragged it into the dirt.

## PERSECUTION GOES ON

Meanwhile, back home, the Wannilaeto were persecuted in innumerable ways. Sometimes they were guilty, and sometimes not. It takes a long while for a traditional hunter and gatherer to see himself as a lawbreaker as he provides for his family like his ancestors did before him. Once, a Wannilaeto man, around sixty years of age, was caught while doing the revolting work of removing the tusks from an elephant in an advanced state of decomposition. Though the park guards saw that the elephant was long dead, the old man was arrested in August 2000. He spent two weeks in jail and was then sentenced to a fine of 10,000 Rupees (about US$131 in the year 2000). On other occasions, there were jeep tracks beside the dead animal. The Wannilaeto do not have a driver's licenses, nor do they drive jeeps, but he was charged nevertheless. In

**Figure 10.10 Tourists Visiting Uru Warige, Wanniya's Home in 2023.** *Source*: W. Stegeborn.

other instances, there were shoe tracks. The Wannilaeto walk barefoot. Their sensitive feet play an important part in their hunting skills. It is essential not to break twigs or walk on dry leaves as they track the game.

The attendant acculturation, however, of the Indigenous culture often results in a breakdown in their traditional values, an "aping" of wholly inappropriate European clothes, technologies, and mannerisms, and again an increase in alcoholism was reported by botanist, zoologist, and biologist Julie Denslow and anthropologist and ethnobotanist C. Padoch (Denslow and Padoch 1988). An undeniable example is Punchi Banda who would still live today had he not worn pants that day in the forest. Western clothes on Indigenous people are obnoxious to park wardens and tourists. When the tourists are displeased, their interest turns to a new destination. (See figure 10.10.)

## HOLIDAY SPENDERS AT WANNILAETO HOME

A dramatic break with the past for the Wannilaeto was to market their modus vivendi to local tourists and foreigners. The almost 23.5 million Sri Lankans often take their holidays to visit the post-hunters and gatherers of the island.

They arrive in cars, buses, and vans, not always sober, with loud music merrily singing the tunes of Bollywood's most popular hits at the time, clapping their hands to the rhythm of the songs. They usually visit during the Sri Lankan New Year in April, during the day of Buddha's enlightenment at the full moon in May, and other holidays. Wannilaeto families; The elderly, women, and children, follow the incursion from their wattle and daub houses, hoping for the best, while their sons and husbands run to the road with handicrafts and forest produce they can sell.

As a rule, the Wannilaeto do not drink alcohol. There are no liquor stores in the rainforest. Bars belong to the pecuniary economy; thus, the few sums of money the post-foragers may have barely cover the most basic food commodities: salt, sugar, and coconuts. The most important reason, however, is that their modus vivendi requires vigilance at all times to avoid scorpions, snakes, and poisonous centipedes on the ground, as well as thieving monkeys breaking corn cobs on the swidden cultivation. Elephants often rampage through their compounds searching for water or cereal grain at night. The family members need to be ready and quick-minded on such occasions. A drunken person is unreliable and considered "mentally unstable" if he acts unbalanced and incoherent. He cannot be trusted as an accountable member of the Wannilaeto societas. Each person, child or adult, has a function in the compound. Hence, when non-Wannilaeto visitors appear bizarre or drunk, the villagers take a step back to observe the person, hoping it will not affect them. With time, however, Wannilaeto men have understood that the jolly crowd wants to befriend them and willingly shares with them their cheese puffs, popcorn, and liquor.

## COMMODIFICATION

One of the most pervasive and far-reaching changes in ethno-tourism is the commodification of people and their environment. It goes right to the heart of the endemic environment. As West et al point out, "It brings new ways of seeing and using people's surroundings to already existing socio-ecological landscapes and creates new boundaries" (West et al. 2006:262). What West et al. are expressing is the right of appropriation. People and Nature are marketed to the rest of the world as products, a view that is alien to the relationships, full of love and respect, that the Wannilaeto share. This benign approach also embraces their intricate socio-ecological landscape, where each mountain has a story, and each forest has its special resources unique to the location and found nowhere else. There is a philosophical and cultural heterogenic difference between the consumers and the consumed. (See figure 10.11)

**Figure 10.11 Elephants and Vultures Scavenge Garbage Outside a Hotel in Polonnaruwa.** *Source*: W. Stegeborn.

The unequal relationship fosters ambiguity in trade regarding the value of sacred places and objects for Indigenous peoples. In addition, the Wannilaeto relationship with their quadruped neighbors changed with the new view of appropriating nature as a commodity. In the Wannilaeto language, for example, there is an expressed affinity between themselves and other beings by making living entities their namesake with the suffix "–aeto." The elk, kankuna-*aeto,* the wild boar, hottedicka-*aeto*, and the snake, pingichcha-*aeto*, all animate, as are the Wannila-*aeto*, the Forest-Beings/Dwellers. Today, however, that relationship is disrupted, and the creatures are considered "park resources" because of their teeth, skin, and claws. Hence, the new generation of Wannilaeto finds their former neighbors better dead than alive.

Government officials directed policies and projects affecting the Wannilaeto without their informed consent. They validated their decisions based on epic stories, religion, and fairy tales. Having created a national park from the trees that were left after the dam constructions, nature and its people became a package deal where the tour guides create simplistic images of the skilled forest people, their customs, and their spiritual understanding of their forest.

Because ecotourists and ethno-tourism penetrate deep into local communities, it causes more damage than other forms of tourism. They bring significant pressure on the Wannilaeto people and natural resources. Destitute and without the means to survive, the Wannilaeto know that tourists desire animal body parts, wild honey, and medicinal plants. They also know that they have

**Figure 10.12 Political Billboard with President Rajapaksha Greeting the Wannilaeto Chief in the Traditional Manner.** *Source*: W. Stegeborn.

to act "fierce and primitive" to meet the tourists' expectations, or the foreign tour guide will not return to them with "good" tourist loads in the future.

Three components are necessary to organize a worthwhile tour: (1) a foreign travel agent, (2) a reliable local counterpart, and (3) a socially competent multilingual tourist guide familiar with European and Sri Lankan languages. With a skilled local travel agent, an uninformed Indigenous community can be persuaded to receive guests, and the agency can offer new "genuine" people in their "natural setting" in their brochures. (See figure 10.12)

## WANNILAETO FOR SALE

Once the Wannilaeto were "discovered," travel agencies and private entrepreneurs organized and promoted tours with Sri Lankan and foreign agents to see the "wild Vedda" people in the jungle. The Wannilaeto became part of a three-day package, where the first day included a visit to an elephant orphanage in Kandy, and the second was a "Night Camp with the Veddahs" where the clients would meet "wild men'" and witness devil possessions. The third day was mountain biking in Polonnaruwa where another eco-reserve was awaiting. Incidentally, if the eco-reserve was void of animals, there were nearby garbage dumps outside the park with its unannounced attraction; elephants, vultures, cows, and dogs scavenging for food. Three to four days before the group arrives, the Wannilaeto disengage from any long-term activities such as long-distance hunting trips because they have to be back by the evening of the arrival of the tourist group. Women do not directly participate in the event, only indirectly by collecting forest resources for trade. The show needed to be small; the more participants, the less the income for each person. The sum was fixed at 2000 Rupees (US$29 in 2000) per performance. Envy and disagreements stirred up as the men competed for the show. Each family needed food, but far from all were lucky.

In anticipation of meeting the visitors, the Wannilaeto gathered items they could sell: medicinal plants, herbs for teas, medicinal wood cups, berries from the forest, dried game meat, and honey. The men strung and tensed their bows and feathered the arrows. Some had carved ebony figurines of elephants, ritual objects, and figurines that represented them as hunter-gatherers. Other desirable items were necklaces of teeth, claws, and strands of hair from an elephant's tail. The latter was supposed to bring luck in a Singhalese or Tamil marriage if the strand was welded inside the gold of a wedding ring. The Wannilaeto do not decorate themselves with dead remains. Nonetheless, they know that people in other parts of the world like to wear them as a sign of personal bravery with an added good story. Lions and tigers do not exist in Sri Lanka, but not everyone knows this. Since the (tiger) tooth necklaces were so highly valued, they did not tell the tourists that, most often, they came from the village dogs.

## NIGHT CAMP WITH THE VEDDAHS

Finally, the day came for the "Night Camp with the Veddahs." On this occasion, the Dutch guide and her local counterpart came twice a month with a group of eco/ethno-tourists. Equipped with white English garden party tents, food, barbecue grills, and coolers with Heineken and Carlsberg beer, they came in buses to befriend the savage "Veddahs." In the evening, placed under the tent roof in case of rain with a beer in hand, the tourists enjoyed the night show of "spirit-possessed" Wannilaeto who performed out in the open, rain or shine. After the show, there was an informal get-together. The guide saw no problem in letting her countrymen, and much more noteworthy, women, offer the Wannilaeto a beer and other stronger beverages after the deliverance. On the contrary, it makes the tourist feel s/he has made contact with an almost undiscovered tribesman, and the guide earns high scores when the tourists evaluate the program at home. Meanwhile, in the Wannilaeto households, worried mothers wondered if their sons would return without trouble, and their wives were concerned about the condition of their husbands when they came home. On the nights when the tourists were treated with displays of "spirit-possessed" dancers, many of them, mostly, young Wannilaeto men became drunk, which shamed their extended families with a "useless" son for a long time thereafter. Not all are easily tempted, however. With a well-calculated plan in mind, Wannilaeto subserviently asked the tourists if they could buy a bottle or two of their fine foreign liquor. Happy to offer something desired by the "jungle brother," some tourists gave a couple of rum or gin bottles or/and whiskeys for free. Some dancers received tips from well-intended tourists to make the spirit-possessed dancers feel appreciated and respected between men and women equals in trade.

After this experience, the guide and her group moved on to new adventures. The Wannilaeto were left to pick up the pieces of garbage, bottles, and half-drunken sons. Half a dozen performers shared 2000 Rupees (approx. US$34 in 1997) for their night's work, including the three to four days of preparations.

When the jeeps, vans, and garden party tents were gone, the real business began. To maximize their earnings, they hurried to pick up the leftovers from the jungle banquet. The most entrepreneurial Wannilaeto grabbed half-emptied liquor bottles and sold them for 500–750 Rupees. Each (Approx. US$ 9–13 in 1997). The unopened bottles of foreign liquor that were obtained from the tourists were sold at triple the price in Mahiyangana. With this money, individuals buy several bottles of arrack that are sold in the village and make everyone happy. The combined eco- and ethno-tourism have expanded. Tourist agents pick up the visitors at the airport or their hotels and bring them to, undoubtedly, spectacular waterfalls, tropical forests, nature walks, exotic animals, and "wild" Indigenous people.

An illustration of how a local student may cause damage to the people he describes is a graduate student from the Hindu College, Tirunelveli in Tamil Nadu, India, who conducted an eighteen-interview field study on the Wannilaeto and tourism in Sri Lanka. He aimed to find whether or not they were marginalized from the mainstream Singhalese society by their own free will (Ranasinghe 2018). He describes programs and projects arranged for the well-being of people such as road building and infrastructure, a government hospital close to Dambana, and the developing of the education sector by building more schools in the areas. He indicates that the majority of the Wannilaeto do not care about the education of their children and adds that "They do not even have proper uniform to wear" (Ranasinghe 2018:11). This is an example of how originating from one's own ethnicity, an assessment is presented based on prejudice and poor preparation of an interview.

The student may not have taken into account that when the Wannilaeto were displaced into the Rehabilitation Villages in the 1980s, the children were molded into the Singhalese former British Crown Colony's educational system. They had to wear English school uniforms with a tie around the neck and laced shoes on their broad, open feet. The girls had to have tightly pleated braids and the boys—short hair. They had to speak in Singhalese, like the female teachers from the capital, not their own Wannilaeto language. The school days started with the Buddhist confession of faith, and the children became adherents of Buddhism. Their parents and grandparents were mocked and ridiculed, and their origin was seen as uncivilized and laughable. This was new to the children. They missed their home environment. At the end of the day, after the children were dropped by the school bus on the main road, they walked 2.5–3 miles to their homes. At that time, there was barely daylight left to be with their mothers and siblings chasing butterflies, whistling bird sounds, and gathering plants, firewood, and tubers. The fathers missed the exploratory rainforest walks with their sons to the mountains while teaching him how to find the way by memorizing map songs. The songs contained landmarks, strange-looking trees, unusual natural formations, creeks, and edible plants that the fathers showed as he sings the monotonous song. The son repeats and one day he can take the lead and walk to the specific location singing the map song while following the landmarks.

Eventually, both parents and children in the Rehabilitation Villages lost motivation to learn the dominant population's interpretation of a "good" life. The young student found through his interviews that the Wannilaeto response to the government's intended improvements in their lives was that they,

> do not care about those facilities given by the government and they do not want survive in the current world. Because they want to protect their culture for future generation [*sic*] without making any changes. (Ranasinghe 2018:11)

Lenient with the government policy of "developing" the Wannilaeto, the student Ranasinghe suggests that potential areas ought to be mapped to attract local and foreign tourists. Public transportation should expand to enhance the growing tourist industry. This would also promote businesses selling forest products. He moreover suggests better marketing of the Wannilaeto cultural rites such as the "Kiri Koraha" (Ranasinghe 2018: 22).

The point in mind is that the Kiri Koraha is a small part of a 24–36-hour religious ceremony held once a year. It is an extracted portion out of its context and lacks meaning to the spectators unless the story, the meaning, and its symbolism are explained to the onlookers. Compare watching the distribution of oblates at communion for a non-Christian. It would not say much.

Ranasinghe emphasizes the need to maximize the Wannilaeto awareness of the importance of education. Education, in this case, would be the norms, rules, and regulations established by both the majority populations, the Singhalese and the Tamils, at the Ministry of Education. Another suggestion is non-formal education for children who drop out of formal schooling.

It is not clear what that is. Maybe it is the home-schooling parents and the extended family carried it out since time immemorial, but it is not referred to as "education" by governments. A final point in his essay is, "Re-allocation of the traditional land rights to the Vedda communities" (Ranasinghe 2018: 22). He considers it beneficial to do this for the preservation of the Wannilaeto culture.

This contradicts the earlier statement about "maximizing the Wannilaeto awareness of the importance of education" (two paragraphs before) (Ranasinghe 2018: 22–28) since state education does not include Wannilaeto culture in its syllabus. The reallocation of traditional lands sounds more like the Wannilaeto perpetual wish to return to their forestland.

At the end of this chapter, there is a comparison of Wannilaeto's conversion process to the mainstream Sri Lankan lifestyle and further adaptation to tourism. This is compared with the experience of the Indigenous people in the Andaman and Nicobar Islands. The recommendations regarding tourism and Indigenous peoples are alike in both Sri Lanka and India.

## SPECIAL OFFER: CHILDREN

Something not yet discussed is the different views on purchasing and consuming women and children in Sri Lanka. In the capital, Colombo, there is a branch of a children's rights NGO with the long name, End Child Prostitution, Child Pornography and Trafficking of Children for Sexual Purposes (ECPAT). They fight an uneven battle against a steady stream of pleasure tourists visiting the west coast and the southeastern parts of Sri Lanka. ECPAT tries to prevent the sexual exploitation of children. Sri Lanka is one of two destinations in Asia specializing in sex tourism. The other is Thailand, where young women/girls are displayed in shop windows and sold as commodities. In Sri Lanka, young boys or youth serve for sodomy by pedophiles and, often, elderly homosexual men. One fortunate condition for parents and children during the COVID-19 pandemic in 2020–2021 was that fewer consumers arrived to harvest the children, although it was a financial loss.

The transition from young Western men and superficially respectable adults in their hemisphere changes as they land at the Bandaranaike airport near the capital in Sri Lanka. Repeat visitors stay in the same hotels, are known by the owners, and have their social networks. Based on earlier years, the hotel owners know their customers' desires and have everything arranged, including the "enjoyment" of children. These are not Wannilaeto children, since they live on the other side of the country. Still, poor fisher families' small ones are children too. With too many mouths to feed, the destitute parents take advantage of the Christian vacations; Christmas, Easter and normal work holidays to offer their son and/or daughter to the hotel owners. The money, "Santa's gift" as the owners call it, provides security for large families when the rainy monsoon is heavy and the catamarans cannot go to sea. Euro-Americans, Chinese, and Japanese men become "Sugar Daddies" year after year until the children are too old and get replaced by younger siblings.

Maybe this can best be illustrated by my earlier driver's extracurricular work as a "fixer." As usual, he picked me up at the airport and brought me to a hotel owned by two male Central European homosexuals who specialize in pedophilia. They cheerfully greeted their old friend, my driver. Being a European, I was expected to embrace his business partners. Jolly and gay men came from restaurants and bars with children and teenage boys who were asked to perform according to the customers' desires. The curtains in my room were wrinkled and deformed textiles of wiped-off human blood and mucus. I never returned.

Over the years, my former driver built up a cache of different commodities for his customers. He lives in a tourist beach town, guides clients around the country, and, depending on their demand, he tries to meet their needs. Many want to see "primitive men," so he brings them to the Wannilaeto. This became quite a rewarding business since he soon built a Bed and Breakfast lodge a few kilometers from Dambana, my "home" village from 1979 onward. He brings tourists from the hotels on the west coast to the Maduru Oya National Park and arranges for the tourists to fraternize with the Indigenous people. The wilder and more primitive he describes the forest people, the braver is the tourist who approaches them. Hence, it can be worthwhile to buy the expensive "tiger tooth" to tell the fellows at the pub at home about the dangerous eye-to-eye encounter.

## EXCLUSIVITY

The more extinct a species becomes, the larger the demand. The request for exotic "bush meat" by well-situated gourmands makes it increasingly challenging for exclusive restaurants in cosmopolitan cities to obtain gorillas, shark fins, and eggs from endangered sea turtles. It is not usually consumed because of its nutritious or palatal qualities, but rather because of its exclusiveness, which few gentlemen can parallel in matters of price and uniqueness. The host gains prestige and high social ratings among his kind.

Similarly, there is a market for Indigenous women. They are also consumed but differently. The young maiden's virtue is that she is a virgin and on the verge of ethnic extinction. Parallel to the consumption of wild meat, the buyer gains prestige among his fellow brothers by describing or documenting his conquests. The likelihood that a man with similar interests beats him is next to none, once the maiden is "popped."

There are numerous international reports on Indigenous women coming home in body bags after their employment as domestic workers. They were raped, sexually harassed, sometimes killed, and organs extracted (Clark and Poucki 2018; Jayakody and Moramudali 2014; Kaye 2017; Moniruzzaman 2006; Moniruzzaman 2012; Prynne 2013; Rasooldeen 2014; Sri-Lanka-Mirror 2014; Vinding 2012). Anthropologist Helena Whall reports on the hunters and gatherers of Sri Lanka who, because of having lost their land and ability to sustain themselves, have entered into the sale of women as domestic aids to other continents.

In Sri Lanka, young Wanniyala-Aetto (Veddha) women who have been removed from their traditional forests and are now living in government Rehabilitation Villages are sold as contract domestic workers to employers in the Middle East for as little as SL Rupees 7,000 per woman. It goes without saying that once abroad, many of these women face sexual abuse from their employers or end up in brothels (Whall, H. 2004).

In the mainly Western novels, the forest dwellers are depicted as good-natured people leading a peaceful and romantic life in the jungle (Knox [1681] 1911; Seligman and Seligman [1911] 1969; Spittel 1924, 1941, 1957; Spittel 1950). The Edenic old images commonly prevail among the Sri Lankans, although the people who profit from the Wannilaeto say they are wild and dangerous. (See figure 10.13.)

Political party representatives, presidents, and prime ministers are people too, and find it a bit tickling to stand face to face with a "Real Veddah." Before the local elections in 2011, they traveled across the country staffed with journalists and photographers. The officials have themselves photographed as

**Figure 10.13 Envelope with drawings of Indigenous people.**

they greet the Wannilaeto in their traditional manner. This demonstrates their bravery and their skill in befriending the so-called wild people of the jungle.

In 1994, the UN launched the International Decade of the World's Indigenous Peoples (1995–2004) and allocated special funds to countries with Indigenous peoples to help them maintain their culture and modus vivendi. Although the Sri Lankan government did not agree with this UN philosophy, the regime saw a way of spreading the word of their exotic people to tourists all over the world. They would imprint the Wannilaeto on stamps! Their image would fly around the globe and bring external revenues to the country at the same time as it illustrated government concern about their "uncivilized" (from the Rotary pamphlet, p. 1) forest people. The UN paid for the advertisement. With the international announcement, the government could expand ethnotourism and "Night Camps with the Veddahs." The Wannilaeto were depicted as ax-holding savages dancing feverishly around a tripod with a cooking pot.

In real life, the former hunters and gatherers continued to be arrested, far from media and UN attention. A wildlife officer killed at least one Wannilaeto man more in Henannegala apart from the earlier mentioned Punchi Banda. The man, Uru Warige Appu's son, Kiribandiya, came from his rice field in System C when he was shot and killed. He was not inside the high-voltage electric-fenced wildlife reserve. No legal action was taken. The park guard is still performing his duties. Almost every month, someone is wounded. In strong contrast to the public image of government concern about the Indigenous people, the Wannilaeto were denied what they most wanted, their former hunting grounds in the forest. Ironically, it was political schemes, through electoral planning, and exploitation of the Wannilaeto in shows and exhibitions, and media that provided a way for the Wannilaeto to speak about their way of life as foragers that now was prohibited. They carefully avoided implicating the government for fear of repercussions, and nothing was said about detaining them for hunting one day and promoting the sale of forest produce to the public the other. The Wannilaeto have learned the mentality of their political representatives.

## COVID-19

The World Health Organization allocated funds to buy the antidote against the coronavirus and its mutations, but there are a series of difficulties. As the Colombo Telegraph says, *Sri Lanka Has Not One, Two Viruses: Covid 19 & Corruption* (Vishwamithra 2021). The public demands transparency. The public also demands equitable access to COVID-19 vaccinations.

On September 4, 2021, Uru Warige, Wanniya's wife Heenmenika, died of a stroke. There were also strains of the Covid-19 virus in her body. She was not returned to her compound, as custom bids. For fear of contamination, she was burned on a funeral pyre in the town of Mahiyangana. With CKD and one kidney transplanted from her daughter, she was not able to withstand all the strains. The link between the Mahaweli Development Program's use of agrochemicals or heavy metals and CKD of unknown etiology remains to be established.

District Medical Officers and health personnel visited Wannilaeto in Dambana twice, at one-month intervals, to vaccinate people over the age of sixty. Medical personnel distributed free soap and instructed everyone to keep an arm-length distance from one another and avoid traditional greetings, that is, resting one's hands on the other person's forearms. Vehicles with loudspeakers patrolled the villages from dawn to dusk, telling people not to leave their homes or let someone enter their houses. The Wannilaeto recognized that they were not supposed to visit the weekly Saturday market in Mahiyangana, 27 km. from there. Everyone was afraid of this new dangerous disease. People who were not inside their households at the time of the first warning by the loudspeakers were not let inside. The old and the feeble died, but with the steady rate of new little Wannilaeto babies, the approximate number of 2,500 Wannilaeto remains unchanged.

## THE SENTINELESE ON THE ANDAMAN AND NICOBAR ISLANDS

Other vulnerable people at the brink of extinction are the different Indigenous peoples on the Andaman and Nicobar Islands. There, the Indian government had undertaken measures to legally protect the Sentinelese population by ensuring them North Sentinel Island territorial sovereignty. They had to be protected against the entry of outsiders, independently of tourism, business entrepreneurs, or explorers (Srivastava VK et al. 2020:172). Having lived there for thousands of years, they were entitled to maintain their traditional and customary rights, the government decided. It was the prime duty of the state to protect these rights as "eternal and sacrosanct," authors affirm (Srivastava VK, et al. 2020:173). With time, statistics showed that these measures gave good results. The Sentinelese population increased (Srivastava VK, et al. 2020:166).

With the 2004 Indian Ocean tsunami, many people on the Great Andaman Islands evacuated to Port Blair, the island group's capital. Their homes were destroyed, and basic services were not in place. Soon thereafter, high up on the hills, to avoid the effects of another tsunami, the local administration built concrete houses on the Strait Island.

Parallel to the Wannilaeto displaced in the Rehabilitation Villages with cement houses and corrugated steel roofs, hot as baking ovens, the islanders of the Strait Island did not show any interest in shifting to those houses as they were away from the seashore (Srivastava VK, et al. 2020:167). Some people remained in Port Blair and adjusted to urban life. Their children went to schools there and were housed in school hostels. But,

> despite improvement in their numbers and economic well-being, something appeared to be missing in the process of development. They did not look to be a happy, thriving population. They still appeared to be devoid of pride, vivacity and élan. (Srivastava VK, et al. 2020:166)

Part of this may be the interactions with local and international tourists. Prostitution, crime, gambling, and drug abuse have been mentioned as the evils of "sun-sea-sand-sex" tourism (Options 2008) in other documents.

In connection to Indigenous people's rights and access to land, the authors of *Draft of the Policies for Great Andamanese and Sentinelese* discuss the splintering of the original *racial stock,* their migration, and similarities and differences which would be of no consequence, the authors say, since they had lived together for a long time and their "rights are unassailable, non-negotiable and uninfringeable" (Srivastava VK, et al. 2020:173). The author's choice of words such as "racial stock" seems untimely year 2020 in the *Journal of the Anthropological Survey of India* in which the article is published, but by looking at the recommendations, it is easy to see the vestiges of Old School Colonial Anthropology.

One of the points is to provide adequate educational facilities in Strait Island, Port Blair or mainland India and ". . . to help the indigenous people in protecting and preserving their cultural heritage and to nurture self-esteem and in their cultural preservation" (Srivastava et al. 2020:167). It should be noted that there are similarities and differences between the Indian government policies vis-à-vis the Andaman Islands and those of the Sri Lankan government vis-à-vis the Wannilaeto.

## SRI LANKAN GOVERNMENT EDUCATING THE WANNILAETO

Sri Lanka did not leave the Wannilaeto to decide for themselves upon their subsistence economy and modus vivendi. The Indian government did, in 1956 (Srivastava et al. 2020). Despite the assertions of Economic, Social, and Cultural rights for Indigenous peoples, both articles recommend state-based education to guide the Indigenous peoples into Sri Lankan or Indian mainstream society, respectively—and to tourism, only more controlled (read: profitable) for the government.

The Wannilaeto can no longer exist on foraging, horticulture, and slash-and-burn cultivation alone. Their land is too scarce, the wildlife too close, and what is left of nature diminishes every day by illicit logging. The only thing they can offer is their exotic appearance. Still, after that, the last tourist bus leaves the showground before dusk at 5 p.m. They return to their own lives, just like any worker. They take off the "wild man's" clothes; a skirt of leaf clad branches, they bathe at the well, put on a sarong, attend to the evening chores, eat rice and curry, and watch the "telly" with family and friends. Something that has changed is the content of their conversations. They no longer tell hunting stories or recount dangerous encounters with wild animals. Now their stories are about stupid "suddas" (white people), money-stringent locals, and how much they gained selling tiger teeth from their toothless dogs. Still, some "Suddas" are good. They paid a lot of money and gave them liquor.

Their cultural change is irrevocable. They will have to carry on acting until the tourists are tired of them and want new amusements, as described in other documents. Many Wannilaeto no longer know who they are. What is a Wannilaeto man? They mimic a fantasy that other cultures have described to them through folk stories, fairy tales, and movies. It is difficult for those who do not want to live by acting. They do not know the codes of conduct in Singhalese, Tamil, or Muslim societies. They do not have the sought-after professional competence to compete for jobs, and they lack contacts and networks to obtain employment.

On the sunny side of eco- and ethno-tourism in Sri Lanka, the Wannilaeto have reached such success in earning money as "wild men" that even Singhalese children speak the Wannilaeto language, sing their songs, and dance like "savages" to attract tourists and earn money. In a twisted way, tourism has contributed to the preservation of the language, the songs, and the mantras (spells against the presence of elephants, snakes, wild boars, and other dangers). They also perform ritual dances even though they do not know where in the authentic ceremonies those dances are conducted nor for which spirit or why. Maybe this is the Wannilaeto solution for survival in the forest for many more years.

Despite the previous three brothers Rajapaksha's governance of Sri Lanka, one as president, another as prime minister, and the third as the minister of Finance, the Wannilaeto life continues as usual. As a result of corruption, nepotism, economic mismanagement, careless borrowing for vanity projects, failure to pay for imports and foreign debts, and the brothers' collective bankruptcy of the country, Sri Lanka now suffers from fuel and food shortages, months of lengthy power blackouts, and record inflation, surpassing 100 percent for some items.

Previously introduced to petrochemical products by the Mahaweli Development Project, the Wannilaeto cannot plant without them. Ukraine was an important exporter of those products to Sri Lanka. Those products can no longer be exported from Ukraine as they are beleaguered by the Russians. The interim Sri Lankan president, Mr. Wickramasinghe, has declared a state of emergency, cracking down on opponents and demonstrations. The twice-a-day national curfew: one in the morning and the other after 6 P.M., affects the Wannilaeto going by bus to town to sell plants and forest produce at the vegetable market in Mahiyangana town. They cannot afford the fines if arrested and cannot pay for an overnight stay at a hotel. In spite of the governmental turmoil, Wannilaeto's life is relatively unchanged. That is the status quo as long as they are not exposed to further outside interference.

## ACKNOWLEDGEMENT

1. I wish to thank the Wannilaeto for generously sharing their knowledge and daily lives with me since 1979. I thank the late "Chief" Uru Warige Tissahamy and his wife U. W. Sudumenika for absorbing me into their world in Dambana. Their eldest son, Wanniya, became my main informant in religion, medicine, survival skills, social behaviors, tactics, strategies, genealogies, and conflict resolution. I also want to extend my gratitude to both my host families, P. G. A. Wijeris (Mudal Ali) and Baby Nona, and subsequently Una Pana Warige, Appu, and his wife Morane Warige, Sudumenika, and their families. Their daughter, U.P.W. Silawatti, was my confidant, helper, and ingenious problem solver from a young age and still is. I am grateful to my friends in the pictures who let me catch a moment in time with them in the foreground: Uru Warige, carpenter "Sudda" (Sudubandiya) and his wife U. W. Karuawattii, Morane Warige, Muttumenika ("Kiri") and her daughter Dammayanti, cutting straw, among others. I also thank anthropologists Robert Hitchcock, with whom I have worked in parallel since the mid-1990s and from whom I continuously receive information on Indigenous peoples, nature reserves, game parks, and human rights. Thayer "Ted" Scudder, a leading authority on large dams and resettlement from the California Institute of Technology, provided valuable information on the Mahaweli Development Project, and archaeologist John Speth of the University of Michigan meticulously read and corrected my dissertation during the first decade of this millennium.

## REFERENCES

Bandaranayake-Kumaratunga, C. 1998. President of Sri Lanka, edited by P. Office, pp. 1, Colombo.

C.I.A. 2023. World Fact Book, Sri Lanka. Office of Public Affairs, Central Intelligence Agency (CIA), Washington, D.C. 20505.

Clark, J. B. and S. Poucki. 2018. *The SAGE Handbook of Human Trafficking and Modern Day Slavery*. SAGE Publications.

Commonwealth, S. 2004. *Gender and Human Rights in the Commonwealth: Some Critical Issues for Action in the Decade 2005-2015*. Commonwealth Secretariat.

Daily-News-Reporter. 1998. Demands of Veddah community to be met.

*Daily News* Wednesday 05, August 1998. Colombo.

Denslow, J. S. and C. Padoch. (eds). 1988. *People of the Tropical Rain Forest*. University of California Press, In association with Smithsonian Institution Traveling Exhibition Service, Washington D.C. Berkeley, Los Angeles, London.

De Zoysa, Mangala. 2021. Forest-Based Ecotourism in Sri Lanka: A Review on State of Governance, Livelihoods, and Forest Conservation Outcomes. *Journal of Sustainable Forestry*, doi: 10.1080/10549811.2021.1943450

Dissanayake, G. 1985. Mahaweli Press Conference w. Dissanayake by Skandinaviska Enskilda Banken. Translated by M. D. M. G. D. j. f. G. R. SEB Bank people Sveriges Radio, Stockholm, Sweden.

Government-of-Sri-Lanka. 2011. Sri Lanka Budget 2012 Salient features. In *New, Sri Lanka*, pp. 4. Department of Government Information, Colombo.

Government-of-Sri-Lanka, T.D.A. 2008. Tourist Arrivals by Country of Residence, Sex and Age group - 2008, edited by T. D. Authority. Government-of-Sri-Lanka, Colombo.

Government, S.-L. 1937. Fauna and Flora Protection (Chapter 567), edited by S. Department, pp. 28. Government of Sri Lanka, Colombo.

IWGIA. 1985. Sri Lanka: Vedda unable to put their case to UN. In *IWGIA Newsletter Nos. 43 & 44 Paperback – 1985*, edited by IWGIA, pp. 289–290, 1 edition,. vol. 43 & 44. IWGIA, Copenhagen, Denmark.

Jayakody, R.L. and U. Moramudali. 2014 Kidney the plot thickens . . . *Ceylon Today* April 28 1. 101 Rosmead Place, Colombo 7: Sri Lanka.

Jayasumana, C., R. Gajanayake and S. Siribaddana. 2014. Importance of Arsenic and pesticides in epidemic chronic kidney disease in Sri Lanka. *Biomedical Center Nephrologi* 15(124): 1–5.

Kaye, J. 2017. *Responding to Human Trafficking: Dispossession, Colonial Violence, and Resistance among Indigenous and Racialized Women*. University of Toronto Press.

Knox, R. (1681) 1911. *An Historical Relation of the Land Ceylon in the East-Indies*. 1911 ed. James MacLehose and Sons, Glasgow, Great Britain.

Lankapuvath-Correspondence. 1999. Two wildlife men arrested for '98 killing of Veddah. *Daily News* May 9:1. Colombo; Sri Lanka.

Madahapola-Ratemahatmaya, N.-W.P. 1973. The Deer and Its Friends. In *Village Folk Tales of Ceylon*, edited by H. Parker, pp. 5–8, 2nd editon, vol. 3. Tisara Prakasakayo, Dutugemunu St., Dehiwala, Sri Lanka.

Mahweli-Environmental-Project 1987 Title., in press.

Moniruzzaman, M. 2006. Moving Human and Non-Human Body Parts: A Review of the History of Organ and Tissue Transplant *The Eastern Anthropologists* 59(2): 179–193.

Moniruzzaman, M. 2012. "Living Cadavers" in Bangladesh: Bioviolence in the Human Organ Bazaar. *Medical Anthropology Quarterly* 26(1): 69–91.

Niyangoda, B.D.J. 1998a. Tissahamy--a Veddah or an impudent imposter? *The Island* June 12. Colombo: Sri Lanka. 1998b. Tissahamy was a Sinhalese. *The Island* July 10. Colombo.

Options, E.T. 2008. *Rethink Tourism in the Andamans - Research Report*. Equations.

Prynne, M. 2013. Britons being sold illegal kidneys for surgery in Sri Lanka. In *Telegraph*. Telegraph Media Group Limited, Contact us by telephone: Telegraph General Switchboard +44 20 7931 2000, 111 Buckingham Palace Road, SW1W 0DT London: Great Britain.

Punchihewa, G. 1989. Souvenirs of a Forgotten Heritage. Colombo: Department of Information, Sri Lanka. Digitized: January 5, 2023.

Rafiq, A.R. 1999 Blue Jeans Vedda falls into a legal trap, claims attorney. *Sunday Times* March 7, 1999. Colombo; Sri Lanka.

Rajapakse, S., M.C. Shivanthan and M. Selvarajah. 2016. Chronic kidney disease of unknown etiology in Sri Lanka. *International Journal of Occupational and Environmental Health* 22(3): 259–264.

Ranasinghe, R.A. 2018. *Marginalization of Indigenous Communities: Evidence from Vedda Indigenous Minority in Sri Lanka*. Uva Wellassa University.

Rasooldeen. 2014. Rising number of housemaid deaths alarms Sri Lanka. In *Arab News*. Arab News. Saudi Research & Publishing Company in the Kingdom of Saudi Arabia. P.O. Box 10452 Jeddah 21433 Saudi Arabia. FAX: +966 2 2836228, Jeddah.

Rotary-Club-Moratuwa. 1990. Tribesmen in Town: A Cultural Show of The Veddahs of Sri Lanka. In *Brocheur*, edited by R. C. o. Moratuwa, pp. 1–4, 1st edition, vol. 20. Rotary Moratuwa, 15/1, Model Av., Model Town, Ratmalana.

Seligman, C.G. and B.Z. Seligman [1911] 1969. *The Veddas*. 1 vols. Cambridge University Press, Cambridge, UK.

Singh, T.V., H.L. Theuns and F.M. Go (eds.) 1989. *Towards Appropriate Tourism: The Case of Developing Countries*. Peter Lang, Frankfurt am Main, Germany.

Spittel, R., Leonald. 1924. *Wild Ceylon* 1. Colombo Apothecaries Co., Colombo.

Spittel, R., Lionel. 1941. *Savage Sanctuary ... Illustrated by Jan Cooke*. London; Colombo Book Centre, Colombo, Sri Lanka. 1957 *Far Off Things,* Second edition, 1. 1 vols. K.V.G. de Silva & Sons, Colombo & Kandy.

Spittel, R.L. 1950. *Vanished Trails, the Last of the Veddas* 1, vol. 1.. Oxford: Oxford University Press.

Sri-Lanka-Mirror, J. 2014. International kidney racket at Thalawathugoda hospital. In *Sri Lanka Mirror,* 1st edition. Sri Lanka Mirror. Chief Editor - Kelum Shivantha, The Think Tank (Pvt) Limited, No. 5/3, 2/1, Railway Avenue, Nugegoda, Sri Lanka.

Srivastava, V.K., Kumar, U., Sasikumar, M., Pulamaghatta, V.N., Patel, S.K. and Goyal, P.A. 2020. Draft of the Policies for Great Andamanese and Sentinelese. *Journal of the Anthropological Survey of India* 69(1): 165–176.

Stegeborn, W., U.W. Wanniya, U.W. Punchi-Banda and U.P.W. Sudubandiya. 1996. UN 1996 Wanniyala-Aetto presentation. Paper presented at the United Nations Working Group on Indigenous Populations, Geneva.

Tampoe, A. 1999. Facsimile transmittal, edited by M. Cardillo, pp. 1. Regarding Punchi Banda's court case ed. Stegeborn, Wiveca, Colombo.

Tampoe, A. 1999. Re. Uruwarige Punchi Banda, edited by A. a. Law, pp. 1. Arun Tampoe, Kotugoda, Colombo.

Vinding, D.K., Ellen Rose. 2012. Indigenous women workers: With case studies from Bangladesh, Nepal and the Americas, pp. 58. The Programme to Promote ILO Convention No. 169 International Labour Office, ILO's International Labour Standards Department 4, Route des Morillons. 1211 Geneva 22, Switzerland.

Vishwamithra. 2021. Sri Lanka Has Not One, Two Viruses: Covid 19 & Corruption. *Colombo Telegraph* June 2, 2021.

Wanigasuriya, K.P., R.J. Peiris-John and R. Wickremasinghe. 2011. Chronic kidney disease of unknown aetiology in Sri Lanka: Is cadmium a likely cause? *Springer Nature: BMC Nephrology* (32).

Weerakon. 1982. President Jayawardene's promise to the Wanniyala-Aetto, edited by President's-Office, pp. 1. Colombo: President's Office..

West, P., J. Igoe and D. Brockington. 2006. Parks and Peoples: The Social Impact of Protected Areas. *Annual Revue of Anthropology* (35): 251–277.

Wijayasiri, L.B. 1998a. Demanads of Veddah community to be met. *Daily News* August 5. Colombo. 1998b President declares Dambana, traditional Veddah area. *Sunday Observer* August 9, 1998. Colombo: Sri Lanka.

Wijesekera, N. 1964. *Veddas in Transition,* 1st edition, vol. 1. M.D. Colombo, Sri Lanka: Gunasena & Co., Ltd..

Wijesinhe, N.D. 1999. Blue Jeans. *Sunday Times* March 14. Colombo.

# Index

# About the Contributors

**F. Sonia Arellano-López** (PhD SUNY Binghamton 2008) is a sociologist whose work has focused on conflicts associated with access to land and natural resources as these express themselves along lines of social class, ethnicity, gender, and age. She has worked for NGOs in South America, as a consultant for international organizations, and as a faculty member in the Institute for Global Cultural Studies at Binghamton University, New York.

**Wayne A. Babchuk** (MA, PhD, University of Nebraska-Lincoln) is a professor of practice in the Quantitative, Qualitative, and Psychometric Methods (QQPM) program in the Department of Educational Psychology at the University of Nebraska-Lincoln (UNL). He serves as a member of the Board of Directors of the Kalahari Peoples Fund (KPF), editor-in-chief for the *Journal of Ethnographic and Qualitative Research* (JEQR), and as an instructor for the African Doctoral Academy at Stellenbosch University, South Africa. Dr. Babchuk is an applied research methodologist who teaches and conducts research on qualitative and mixed methods research across disciplines, research ethics, grounded theory, ethnography, grounded theory ethnography, mixed methods-grounded theory, and community-based participatory research. He is also involved in several other research tracks, including Kalahari San land use and resource rights, conservation and development, ecotourism, applied and development anthropology, the evaluation and assessment of teaching among faculty in post-secondary institutions, student and instructor perceptions of the impact of social media on learning, interdisciplinary collaboration, stalking victimization among sexual and gender minority individuals, and several other topics. He is currently co-authoring two texts on qualitative and mixed methods research and one on ecotourism among Indigenous groups.

**Megan Biesele** is a cultural anthropologist who received her PhD in anthropology from Harvard University in 1975. She has since researched and worked on Ju|'hoan San folklore, belief systems and healing, land rights, mother-tongue education, and language documentation in Botswana and Namibia. She has taught anthropology at the University of Texas, Rice University, Texas A&M University, and the University of Cape Town. She co-founded, in 1973, with Harvard colleagues, a US nonprofit advocacy group, the Kalahari Peoples Fund, and continued as its director until 2023. Among her publications are three books and many articles on Ju|'hoan folklore.

**Roderick Taufan Jan Buiskool** is a development and tourism practitioner. He received his Master of Arts in Development Studies at the University of Maastricht, the Netherlands, in 2020. He has conducted research on socioeconomic issues, specifically on community-based ecotourism in areas bordering the Mount Leuser National Park in North Sumatra, Indonesia, as well as conservation issues and human–wildlife interactions in Way Kambas National Park, Southern Sumatra, Indonesia. He also has worked with United Nations agencies focusing on socioeconomic research in Timor-Leste. Currently, he is the North Sumatra representative of PUM Netherlands senior experts, a Dutch development organization providing advice for small and medium enterprises, as well as coaching for entrepreneurs in developing countries.

**Vidal Gonzales** is from Santa Clara Pueblo. He received his master's degree in Community and Regional Planning with a concentration in Indigenous Planning from the University of New Mexico in the spring of 2023. While working toward his master's degree, he studied the built environment of his Puebloan people and their ancestors. Mr. Gonzales has worked for the National Park Service for over seven seasons, helping to run an Indigenous pueblo youth corps program that works to preserve their ancestral ruins in Bandelier National Monument. Vidal Gonzales now works for the North Central New Mexico Economic Development District as a Community Development Planner to help communities, tribes, and local governments with their community projects and needs.

**Robert K. Hitchcock** (BA, UC Santa Barbara; MA, PhD, University of New Mexico) is a professor in the Department of Anthropology at the University of New Mexico, USA. He is also a board member of the Kalahari Peoples Fund (KPF), a non-profit 501(c)(3) organization that provides funding for education, development, and capacity-building training for Indigenous and minority peoples in southern Africa. He has taught at South Dakota State University, the University of Nebraska-Lincoln, Michigan State University, Truman State University, and the University of New Mexico. Much of his professional career has been spent working on human rights issues facing current and former hunters and gatherers and agropastoral peoples, particularly in southern and eastern Africa. He has worked for several African governments (Botswana, Eswatini, Namibia, and Somalia), and he has served as a consultant for the United Nations Development Program, the United States Agency for International Development, the World Bank, the United Nations High Commissioner for Refugees, and several international non-governmental organizations, including the International Work Group for Indigenous Affairs.

**Melinda C. Kelly** (BA, University of Michigan) has worked on San and wildlife and tourism issues in Botswana and Namibia since 1975, when she was a member of the University of New Mexico Kalahari Project. On her return to the United States, she had a long career as a travel agent, traveling to six continents, while maintaining a strong interest in Africa and the field of anthropology. Her research has focused on human–wildlife interactions, women, craft production, ethnohistory, and Indigenous knowledge. Currently, she is working with the Kalahari Peoples Fund (KPF) on issues involving Indigenous people in southern Africa. She has published numerous papers on topics relating to Indigenous peoples in southern Africa, including those in Botswana, Namibia, and Zimbabwe.

**Stasja Koot** (MA Utrecht, PhD University of Tilburg, Leiden) is an associate professor at the University of Wageningen, Sociology of Development and Change Group, in the Netherlands and a senior research fellow at the University of Johannesburg, Department of Geography, Environmental Management and Energy Studies. He has been working with Indigenous groups in southern Africa, predominantly Namibia and South Africa, since the late 1990s as a researcher and a practitioner in community-based tourism. Recently, his geographical focus has expanded to Indonesia. As an environmental anthropologist, his core focus is on political ecology with an emphasis on power dynamics in tourism and nature conservation. Moreover, he has published about and/or currently works on capitalism, land grabbing, philanthropy, autoethnography, trophy hunting, branding, race and ethnicity, wildlife crime, gated communities, belonging, and development.

**Sachiko Kubota** (PhD Konan University, Kobe 2000) is a cultural anthropologist who is presently the president of Ashiya University in Japan and Professor Emeritus at Kobe University. She has taught at Otemae University, Kobe University, and Hiroshima University. Much of her work has been on *Yolngu*, Australian Aboriginals in northern Australia. Her research interest includes gender issues, social change and Aboriginal arts and crafts. Recently she has been running a new project on the change of the relationship between Indigenous and the nation states from a comparative point ot view.

**R. Fleming Puckett** (DPhil, Oxford University) is a cultural geographer, lawyer, and national meeting speaker who is a board member of the Kalahari Peoples Fund (KPF). At KPF, Fleming is working to increase awareness, funding, and advocacy related to land rights, community governance, and San/government relationships across southern Africa. Much of his work has been among the ǂKhomani San of South Africa, along with the !Xun and Khwe San, who are settled near Kimberley. At the Education Advisory Board (EAB), Fleming's speaking work centers on the power of private philanthropy to expand academic programs and increase access to higher education. His ongoing Kalahari research and publications (begun in 2008) explore the ways in which Indigenous systems of governance, decision-making, land use, and community organization may conflict with the stated goals, requirements, and implementation procedures of national land and "development" legislation and how such conflicts hinder the efforts of Indigenous groups as they seek land, recognition, human rights, socioeconomic improvements, livelihood options, water and food security, political empowerment, and their own senses of cultural preservation.

**Fredy Rodriguez-Mejia** (BA, University of Delaware; MA, Northern Arizona University; PhD, Michigan State University) is currently a faculty member in the Arts & Sciences Division of Northwest State Community College in Archbold, Ohio. He is also a researcher in the Mechanical Engineering Education Research Center at Purdue University. He has taught at Michigan State University, Augustana College in Rock Island, Illinois, and Illinois State University. Much of his work is on gender, identity, tourism, and indigeneity among the Ch'orti' Maya of Honduras, where he has done extensive research and outreach work.

**Wiveca Stegeborn** (BA, University of Uppsala, Sweden; MA, Washington State University) is a cultural anthropologist and advocate for Indigenous peoples in South Asia. Based on her research from 1979 onward, she has provided information on the last 2,500 Wannilaeto to the United Nations. Since 1985, she has presented the Wannilaeto case in other human rights forums. Her work has significantly contributed to raising awareness about the challenges confronted by this hunting and gathering community in the face of modernization, economic development, nature conservation efforts, and violations of their Indigenous rights. She speaks their language and the majority language of Sri Lanka, Singhalese, and has assisted the Wannilaeto petitions to the government to allow them to remain in the rainforest instead of being displaced into Rehabilitation Villages as intensive agriculturalists. She has contributed to three encyclopedias, the latest in 2021. Her most recent fieldwork in Sri Lanka was carried out in 2023, where she witnessed substantial changes and negative social and health impacts among the Wannilaeto.

www.ingramcontent.com/pod-product-compliance
Ingram Content Group UK Ltd.
Pitfield, Milton Keynes, MK11 3LW, UK
UKHW020840290926
15470UKWH00004B/1458

9 781793 6438